MINDSTORMS

Frontispiece: LOGO Turtle.

MINDSTORMS

Children, Computers, and Powerful Ideas

SECOND EDITION

SEYMOUR PAPERT

BASIC BOOKS

A Member of the Perseus Books Group

Designed by Vincent Torre

Library of Congress Cataloging-in-Publication Data

Papert, Seymour.
    Mindstorms: children, computers, and powerful
    ideas.

Includes bibliographical references and index.
1. Education.   2. Psychology.
3. Mathematics—Computer-assisted instruction.
I. Title.
QA20.C65P36 1980 372.7   79–5200
ISBN: 0–465–04627–4 (cloth)
ISBN: 0–465–04629–0 (paper)
ISBN: 0–465–04674–6 (second edition paper)

Contents

Foreword to the Second Edition by John Sculley

WHEN SEYMOUR PAPERT was writing the first edition of *Mindstorms* back in the late 1970s, Apple Computer had just introduced the Apple II. It had 4k of standard memory, and customers used their own TV sets as monitors. At that time, there were 200,000 personal computers in the world, and if people wanted to use them, most had to go through some kind of basic—and usually painstaking—instruction.

Today, over 200,000 personal computers are manufactured in a single week. More important, personal computers play a central role in our lives and have become indispensable in the educational environment. Although the pace and magnitude of change in the industry since 1980 have been nothing short of revolutionary, Seymour Papert's insights and observations about children, computers, and computer cultures remain timeless. In fact, in the current climate of massive educational restructuring, Papert's message is more relevant than ever.

During the Industrial Age and for most of this century, America stood alone at the top of an economic pyramid, taking resources out of the ground—oil, wheat, and coal—adding its manufacturing know-how to those resources, and selling those goods to the rest of the world. We are no longer in the Industrial Age. We are in an Information Economy where strategic advantage is determined by ideas and information and by the skills of a nation's work force.

Virtually overnight, America has gone from being resource rich to being resource poor. As a direct consequence, America is perched on the edge of an economic cliff, and unless we make a concerted effort to bring the educational system into sync with the rest of the global economy, we are in danger of supplying the rest of the world with low-wage work and losing out on the high-skill, high-wage economy that the rest of the world has moved toward.

Seymour was among the first to see that massive change was needed in the education system, particularly math and science education, and to recognize the role that technology could play in learning. Perhaps more significant, though, was how he was one of the first to recognize that technology in the classroom was not a silver bullet that would solve all of education's ills. He realized, long before the rest of the education reform movement, that technology in education is effective only if placed in a larger context that combines well-prepared teachers with integrated social services.

It comes as no surprise that this was similar to the basic philosophy that drove the development of the first Apple computer. Like the group who founded Apple, Seymour Papert saw the potential of a computer to transform fundamentally the way people think, work, learn, and communicate. And the two had something else in common: They acknowledged the amazing intuitive ability of children between the ages of one and three to absorb information in massive amounts that are unequaled at any other time in their lives; they shared an equal amazement for how a child can learn two languages without confusing them; and, finally, they acknowledged the fact that although adults think they can teach children their experience, there is an enormous amount that adults can learn from them. It is worth recalling that it was through research on children that pioneers in human interface design came up with the mouse pointing device.

The two central themes of the book—that children can learn to use computers in a masterful way and that learning to use computers can change the way they learn everything else—were similarly the basis for the Vivarium project that Apple Fellow Alan

Kay headed in the early 1980s. Even outside the classroom, both Alan Kay and Seymour Papert shared a similar vision of the computer being used just as casually and as personally for a diversity of purposes throughout a person's entire life. The passage of time has proved them to have been both prescient and remarkably accurate in their observations of the role personal computing devices will play in our lives.

An important finding of our Apple Classroom of Tomorrow (ACOT) research, and a point that Seymour makes, is that the introduction of computer technology into the classroom did not have the expected—alienating—effect that many people worried about back in the 1970s. On the contrary, what we have found is that in classrooms saturated with technology there is actually *more* socialization and that the technology often contributes to greater interaction among students and among students and instructors.

And so it makes sense to revisit the time-tested ideas that Seymour advanced over a decade ago and to reflect once again on the powerful role that technology can have when placed in the hands of individuals. As experience has shown, the thoughtful introduction of information technology into a social system can fundamentally transform the way intelligence is developed and the way people live their lives.

If the brief history of the personal computing industry has taught us anything, it is that the past is a prelude to what we can expect from information technology devices in the future. It has also demonstrated that often the more counterintuitive propositions are the ones that quickly become part of the collected wisdom of our age. So it is with added enthusiasm that I look forward to rereading the sometimes contrarian—now cherished—ideas of Professor Papert and look forward to his future contributions, which will undoubtedly influence the future course of technology in education.

John Sculley
Chairman & CEO
Apple Computer, Inc.

Foreword to the Second Edition
by Carol Sperry

> Only rarely does some exceptional event lead people to reorganize their intellectual self-image in such a way as to open up new perspectives on what is learnable.
> —Seymour Papert

I AM FAIRLY certain that Seymour Papert had no idea when he wrote the above sentence that the "exceptional event" for many people, for many teachers, would be the appearance of his 1980 book *Mindstorms*. I was one of those teachers, and it is difficult to express the path of my subsequent evolution without sounding a bit melodramatic. I am quite sure, however, that many teachers have had similar experiences. I know this is true because in the past twelve years I have met and worked with teachers from all over the world—throughout the United States, Lithuania, Russia, Costa Rica, Japan—who have been influenced profoundly by *Mindstorms*.

In the spring of 1981, I signed up for a computer workshop at the New York Academy of Sciences. By the time the August workshop was to begin, I could not imagine why I had done such a foolish thing. I felt like a walking stereotype—math, science, and technology phobic, as well as confused by the many contradictions found in the bureaucratic structure of the educational system. The motivating force for my participation in the workshop was my fifth- and sixth-grade students and their interest in computers. Just as I occasionally tuned into their favorite television programs, I wanted to know more about this new great passion. But I felt anxious about my own capabilities and distrusted anything mechanical. I actually played hooky the first day of the workshop.

Guilt drove me out on the second day, and I stepped into a different world. The workshop, organized by Papert, was very much like this book: a dynamic mixture of heady ideas, down-to-earth exercises, and engrossing anecdotes that gave shape to the abstractions of the intellect and breathed life into them. Another surprise: We were all having a good time—instructors and participants alike; we were having fun. I learned the computer language Logo and heard the philosophical ideas that influenced the technical building of the language. My feelings surprised me as I learned to negotiate my way through the language and mathematics of Logo and the mechanics of the computer. I was astonished to hear Papert and his colleagues validate what I had always felt were my best ways of learning and teaching. Here were people who understood the value of process and believed in the construction of knowledge.

During this time, I began to read *Mindstorms,* and from the very beginning my teaching life, which I believe is inseparable from who I am, was deeply affected. After reading the beautifully wrought preface where we learn about the toddler in love with gears, we want to say, "Yes, that's right. That's what learning really is." We recognize the bizarreness of the notion that standardized testing can tell us anything about what children understand and know. I puzzled over why beginning to use Logo could so liberate my constricted feelings about learning and could help so easily reintroduce me to myself as learner, helping me experience again what that could feel like. My connection to my students felt stronger than ever. I knew that only by joining in this adventure with learning could I one day share it with my students. I found myself wondering if this language and this technology could be used, self-consciously, as a vehicle to help build such attributes as self-esteem and self-empowerment in teachers as well as in children.

There are simple things about the workings of Logo that woke me up, that captured my imagination, that made me smile. I liked the idea of doodling in a "command center," sending the turtle/cursor this way and that way, making "serious" shapes (like squares and circles) and not-so-serious shapes (like abstract drawings). When finished, I could capture most of these and felt excitement upon realizing I could name the procedure anything I wanted and later saw that same excitement in my students when they called their shapes by their own names or by a combination of letters meaning something to only them. This made the whole notion of symbols and naming a lesson in the tyranny, or at least the rigidity, of the one correct answer. It is rather shocking to our deeply held metaphysical beliefs to find that names are rather arbitrary, that a name can be disassociated from what it usually means yet the shape of the so-named object does not change. Imagine the children's delight when discovering this new truth and what it means to their deepest understanding of language and laws. When Logo is used well—that is, when the teacher and students use it in their own creative ways—all the boundaries are broken: Children are engaged in activities that are meaningful to them; they are actually talking in school, talking in order to learn, talking to share their discoveries or to ask their classmates for advice.

One of the most powerful aspects of *Mindstorms* is Papert's ability to express complex affective and cognitive ideas in a manner that resonates profoundly with teachers' own deepest feelings regarding their self-worth, their abilities to learn and facilitate learning, and their confidence in their own instincts. While it is common to say in exploratory situations that there are no correct answers, Papert's thinking about learning, made tangible through Logo, makes palpable the philosophies where many truths take the place of one confining truth. This gives us a way to be more expansive in our teaching, to allow for many perspectives, and to honor diverse learning styles. This way of thinking, which is to a fixed curriculum what masterful painting is to painting by numbers, is not without its darker, unnerving side. The teacher must face the exhilarating (though scary) responsibility of forging her own philosophy of education. And since the authentic teacher lives her philosophy, this is inseparable from what Catherine Bateson calls "composing a life." Of course, those teachers who constantly struggle to reconstruct their own knowledge in the course of helping students construct theirs have always accepted the responsibility. What Logo offers them is a new support in an area where few exist. But it can do so only if, so to speak, we have the courage to put ourselves into the disk drive with Logo.

Carol Sperry
Harvard Graduate School of Education

Introduction to the Second Edition

A CENTRAL THEME of *Mindstorms* is that people seldom get anything exactly right on the first try. Intellectual activity does not progress, as logicians and the designers of school curricula might want us to believe, by going step-by-step from one clearly stated and well-confirmed truth to the next. On the contrary, the constant need for course corrections, which I call "debugging" in this book, is the essence of intellectual activity.

Having stated this view so emphatically, I would be embarrassed if I did not use the opportunity of a new edition to comment on "bugs" in the original. I begin by noting some that have more to do with the presentation of ideas than with the ideas themselves. Most of these problems had their origin in a mistaken model of the book's public. In my 1993 book, *The Children's Machine: Rethinking School in the Age of the Computer,* I comment on how much I learned because so many teachers, especially elementary school teachers, not only read *Mindstorms* but also exercised real ingenuity in adapting the realities of life in contemporary schools to ideas that I imagined as having their application only in a more futuristic setting. My failure to anticipate the extent to which the book would be picked up by teachers shows itself most clearly in the gratuitous use of such examples as Newton's laws of motion to illustrate ideas that could as well have been brought out through situations closer to their interests.

As a result, many teachers stopped reading when they hit chapter 5, which bristles with such stuff. By that time many of them had read enough to develop for themselves the ideas that were still to come. However, I think that everyone who likes the book at all would enjoy some of the later parts, particularly chapter 8, entitled "Images of the Learning Society." In fact, I would now recommend reading that chapter first, as one would peep at the end of a mystery story. Doing so will help give readers a firmer idea from the beginning of where the argument is going.

I am not sure whether it was my bug or theirs that led academic critics to rush into experiments to prove me wrong in predicting that "doing Logo" or "working with computers" would *cause* change in how children think. I make no such claim anywhere in the book, but I may have made a mistake in waiting until chapter 8 before saying emphatically that I was not making it. What I was saying, and still say, is something slightly more subtle: I see Logo as a *means* that *can, in principle,* be used by educators to *support the development* of new ways of thinking and learning. However, Logo does not in itself produce good learning any more than paint produces good art.

I have been extraordinarily gratified to see teachers using Logo as a painter uses paint, that is, as a medium for creative work, in this case for the creation or enhancement of learning environments. How they do this is extremely varied, and the meaning given by their students to what they do is even more varied. For this reason, and for many others as well, there is no sense to what I call the technocentric questions that have been echoed in the titles of countless articles in popular and scholarly publications: What is the effect of Logo (or of programming or of computers) on how students think (or learn math or learn to spell or whatever)? Very different phenomena will be seen depending on what is done with Logo—and even when everything else is the same, different students will react in different ways.

What kinds of differences are relevant? From the perspective of the 1990s, it appears bizarre or downright reactionary that *Mindstorms* makes no reference to gender or multiculturalism. I have become convinced that recognizing the androcentric nature of traditional ways of knowing will play a central role in producing change in education. A political reason for this conviction is feminism's strength as a potential ally of other forces working for deep change in education. A more conceptual reason is the belief that feminist epistemological studies have so far provided the deepest sources of insight into human differences. However, the ultimate theoretical task in advancing, for example, the learning of mathematics, is not producing a range of so-and-so-centric kinds of mathematical knowing but rather finding ways of thinking about mathematical knowledge that will allow each individual to make what in *Mindstorms* I call a syntonic appropriation. Thus, although in more recent writing I have sought to ally myself intellectually with contemporary trends toward alternative epistemologies, I remain ambivalent about whether to think of *Mindstorms'* position in this respect as a bug or a virtue.

Mindstorms unquestionably has a bug for giving prominence to structured programming as a model for thinking about thinking. I could say, in mitigation, that I occasionally point out that this is just one model and even propose to use it to define "less mechanical" ways of thinking by contrast. Nevertheless, I can understand why the book seems to have had for some readers a net effect of strengthening the tendency to see structured, analytical thinking as synonymous with good thinking that is inherent in computer science, in educational theory, and indeed in the "canonical" traditional epistemology itself. A reason for this might be that although *Mindstorms* emphatically proposes the idea of "bricolage" as a model for general scientific theorizing, this idea comes late in the book and is not developed as an alternative style of programming; I was more explicit about this in later writing, including, especially, my collaborative work with Sherry Turkle. Thus, teachers who approached the book for advice on how to use Logo and dropped it when they encountered the above-mentioned rough spots could not be faulted for going away with the idea of teaching their students to use only structured programming.

This narrow view of programming was encouraged by the material conditions under which teachers tried to apply the ideas of

Mindstorms and played a significant role in molding the image of programming that formed in the school computer culture. During the 1970s, we had demonstrated that children of almost any age could learn to program in Logo under good conditions with plenty of time and powerful research computers. Today, accumulated experience and technical advances allow even better work to be done with machines that schools can afford to buy in large quantities. However, the conditions in typical schools in the early 1980s were very different: The commercial versions of Logo that came out for the school computers of that time were sufficient to cause great excitement but fell far, far short of what could be done in the research laboratories of the 1970s or the elementary schools of the 1990s. Ironically, I believe that computation in schools might be far more advanced today if it had been kept in abeyance for ten years. Learning to program in the early versions was considerably harder, and much less could be done with the programming skill once it was learned.

To make matters worse, the small number of machines further restricted the levels of programming skill that came to be considered standard. Thus, a primitive model of "programming" became established in schools and was built into large segments of the school computer culture. In *Mindstorms* I introduce Logo programming through examples such as writing tiny programs to make a square and a triangle and putting them together to form a house. Unfortunately, the conditions of work in the schools often meant that what was described in the book as a tiny first step was all that could be done. Under these conditions programming had to be "structured" to do anything at all. Moreover, as other "prepackaged" uses of the computer became more abundant, many educators wondered whether programming was worth the trouble.

The idea of programming for children in schools was kept alive by two circumstances. The first was the emergence of a cohort of visionary Logo teachers. In *The Children's Machine* I describe how imaginative progressive teachers who had computers in their classrooms and were prepared to give students the time and support to learn often created wonderfully fertile environments where difficulty was a challenge rather than an obstacle. On the negative side, I also describe a trend that became dominant in most schools: moving computers into "computer labs" where a routinized curriculum of "computer literacy" would be implemented. Under these conditions, learning often inherited all that was worst in curriculum-driven school practices and learning anything as "difficult" as "programming" seemed to be would have gradually withered away—without the second saving circumstance.

As machines have become more powerful, the programmers who implement versions of Logo have become more skillful and ideas about what children could do through programming have deepened. The result has been a succession of versions of Logo—the most significant steps were LogoWriter, which appeared in late 1980s; Lego-Logo, which followed soon after; and most powerfully, Microworld Logo, whose publication will be simultaneous with this book. Each of these steps lowered the effort of learning to program and raised the interest and complexity of what could be done with a given level of skill. At the same time, the number of computers in schools grew steadily so that more time could be spent on programming in those places where the idea had survived the bleaker period. Most important of all, in many schools students were now able to use programming as an expressive medium to study other topics rather than as a skill to be learned for the sake of learning it. As they do so they become fluent, and as they become fluent they begin to use their own varied styles of programming. Thus, history is curing the worst of the bugs I recognize in *Mindstorms* and perhaps opening these ideas to new phases of development.

Preface

The Gears of My Childhood

BEFORE I WAS two years old I had developed an intense involvement with automobiles. The names of car parts made up a very substantial portion of my vocabulary: I was particularly proud of knowing about the parts of the transmission system, the gearbox, and most especially the differential. It was, of course, many years later before I understood how gears work; but once I did, playing with gears became a favorite pastime. I loved rotating circular objects against one another in gearlike motions and, naturally, my first "erector set" project was a crude gear system.

I became adept at turning wheels in my head and at making chains of cause and effect: "This one turns this way so that must turn that way so . . ." I found particular pleasure in such systems as the differential gear, which does not follow a simple linear chain of causality since the motion in the transmission shaft can be distributed in many different ways to the two wheels depending on what resistance they encounter. I remember quite vividly my excitement at discovering that a system could be lawful and completely comprehensible without being rigidly deterministic.

I believe that working with differentials did more for my mathematical development than anything I was taught in elementary school. Gears, serving as models, carried many otherwise abstract ideas into my head. I clearly remember two examples from school math. I saw multiplication tables as gears, and my first brush with equations in two variables (e.g., $3x + 4y = 10$) immediately evoked the differential. By the time I had made a mental gear model of the relation between x and y, figuring how many teeth each gear needed, the equation had become a comfortable friend.

Many years later when I read Piaget this incident served me as a model for his notion of assimilation, except I was immediately struck by the fact that his discussion does not do full justice to his own idea. He talks almost entirely about cognitive aspects of assimilation. But there is also an affective component. Assimilating equations to gears certainly is a powerful way to bring old knowledge to bear on a new object. But it does more as well. I am sure that such assimilations helped to endow mathematics, for me, with a positive affective tone that can be traced back to my infantile experiences with cars. I believe Piaget really agrees. As I came to know him personally I understood that his neglect of the affective comes more from a modest sense that little is known about it than from an arrogant sense of its irrelevance. But let me return to my childhood.

One day I was surprised to discover that some adults—even *most* adults—did not understand or even care about the magic of the gears. I no longer think much about gears, but I have never turned away from the questions that started with that discovery: How could what was so simple for me be incomprehensible to other people? My proud father suggested "being clever" as an explanation. But I was painfully aware that some people who could not understand the differential could easily do things I found much more difficult. Slowly I began to formulate what I still consider the fundamental fact about learning: Anything is easy if you can assimilate it to your collection of models. If you can't, anything can be painfully difficult. Here too I was developing a way of thinking that would be resonant with Piaget's. *The understanding of learning must be genetic.* It must refer to the genesis of knowledge. What an individual can learn, and how he learns it, depends on what models he has available. This raises, recursively, the question of how he learned these models. Thus the "laws of learning" must be about how intellectual structures grow out of one another and about how, in the process, they acquire both logical and emotional form.

This book is an exercise in an applied genetic epistemology expanded beyond Piaget's cognitive emphasis to include a concern with the affective. It develops a new perspective for education research focused on creating the conditions under which intellectual models will take root. For the last two decades this is what I have been trying to do. And in doing so I find myself frequently reminded of several aspects of my encounter with the differential gear. First, I remember that no one told me to learn about differential gears. Second, I remember that there was *feeling, love,* as well as understanding in my relationship with gears. Third, I remember that my first encounter with them was in my second year. If any "scientific" educational psychologist had tried to "measure" the effects of this encounter, he would probably have failed. It had profound consequences but, I conjecture, only very many years later. A "pre- and post-" test at age two would have missed them.

Piaget's work gave me a new framework for looking at the gears of my childhood. The gear can be used to illustrate many powerful "advanced" mathematical ideas, such as groups or relative motion. But it does more than this. As well as connecting with the formal knowledge of mathematics, it also connects with the "body knowledge," the sensormotor schemata of a child. You can *be* the gear, you can understand how it turns by projecting yourself into its place and turning with it. It is this double relationship—both abstract and sensory—that gives the gear the power to carry powerful mathematics into the mind. In a terminology I shall develop in later chapters, the gear acts here as a *transitional object.*

A modern-day Montessori might propose, if convinced by my story, to create a gear set for children. Thus every child might have the experience I had. But to hope for this would be to miss the essence of the story. *I fell in love with the gears.* This is something that cannot be reduced to purely "cognitive" terms. Something very personal happened, and one cannot assume that it would be repeated for other children in exactly the same form.

My thesis could be summarized as: What the gears cannot do

the computer might. The computer is the Proteus of machines. Its essence is its universality, its power to simulate. Because it can take on a thousand forms and can serve a thousand functions, it can appeal to a thousand tastes. This book is the result of my own attempts over the past decade to turn computers into instruments flexible enough so that many children can each create for themselves something like what the gears were for me.

MINDSTORMS

Introduction

Computers for Children

JUST A FEW YEARS AGO people thought of computers as expensive and exotic devices. Their commercial and industrial uses affected ordinary people, but hardly anyone expected computers to become part of day-to-day life. This view has changed dramatically and rapidly as the public has come to accept the reality of the personal computer, small and inexpensive enough to take its place in every living room or even in every breast pocket. The appearance of the first rather primitive machines in this class was enough to catch the imagination of journalists and produce a rash of speculative articles about life in the computer-rich world to come. The main subject of these articles was what people will be able to do with their computers. Most writers emphasized using computers for games, entertainment, income tax, electronic mail, shopping, and banking. A few talked about the computer as a teaching machine.

This book too poses the question of what will be done with personal computers, but in a very different way. I shall be talking about how computers may affect the way people think and learn. I begin to characterize my perspective by noting a distinction between two ways computers might enhance thinking and change patterns of access to knowledge.

Instrumental uses of the computer to help people think have

MINDSTORMS

been dramatized in science fiction. For example, as millions of "Star Trek" fans know, the starship *Enterprise* has a computer that gives rapid and accurate answers to complex questions posed to it. But no attempt is made in "Star Trek" to suggest that the human characters aboard think in ways very different from the manner in which people in the twentieth century think. Contact with the computer has not, as far as we are allowed to see in these episodes, changed how these people think about themselves or how they approach problems. In this book I discuss ways in which the computer presence could contribute to mental processes not only instrumentally but in more essential, conceptual ways, influencing how people think even when they are far removed from physical contact with a computer (just as the gears shaped my understanding of algebra although they were not physically present in the math class). It is about an end to the culture that makes science and technology alien to the vast majority of people. Many cultural barriers impede children from making scientific knowledge their own. Among these barriers the most visible are the physically brutal effects of deprivation and isolation. Other barriers are more political. Many children who grow up in our cities are surrounded by the artifacts of science but have good reason to see them as belonging to "the others"; in many cases they are perceived as belonging to the social enemy. Still other obstacles are more abstract, though ultimately of the same nature. Most branches of the most sophisticated modern culture of Europe and the United States are so deeply "mathophobic" that many privileged children are as effectively (if more gently) kept from appropriating science as their own. In my vision, space-age objects, in the form of small computers, will cross these cultural barriers to enter the private worlds of children everywhere. They will do so not as mere physical objects. This book is about how computers can be carriers of powerful ideas and of the seeds of cultural change, how they can help people form new relationships with knowledge that cut across the traditional lines separating humanities from sciences and knowledge of the self from both of these. It is about using computers to challenge current beliefs about who can understand what and at what age. It is about using computers to question standard assumptions in developmental psychology and in the psychology of aptitudes and attitudes. It is about whether personal computers and the cultures in which they are used will continue to be the creatures of "engineers" alone or whether we can construct intellectual environments in which people who today think of themselves as "humanists" will feel part of, not alienated from, the process of constructing computational cultures.

But there is a world of difference between what computers can do and what society will choose to do with them. Society has many ways to resist fundamental and threatening change. Thus, this book is about facing choices that are ultimately political. It looks at some of the forces of change and of reaction to those forces that are called into play as the computer presence begins to enter the politically charged world of education.

Much of the book is devoted to building up images of the role of the computer very different from current stereotypes. All of us, professionals as well as laymen, must consciously break the habits we bring to thinking about the computer. Computation is in its infancy. It is hard to think about computers of the future without projecting onto them the properties and the limitations of those we think we know today. And nowhere is this more true than in imagining how computers can enter the world of education. It is not true to say that the image of a child's relationship with a computer I shall develop here goes far beyond what is common in today's schools. My image does not go beyond: It goes in the opposite direction.

In many schools today, the phrase "computer-aided instruction" means making the computer teach the child. One might say the *computer is being used to program* the child. In my vision, *the child programs the computer* and, in doing so, both acquires a sense of mastery over a piece of the most modern and powerful technology and establishes an intimate contact with some of the deepest ideas from science, from mathematics, and from the art of intellectual model building.

I shall describe learning paths that have led hundreds of children to becoming quite sophisticated programmers. Once programming is seen in the proper perspective, there is nothing very surprising about the fact that this should happen. Programming a computer means nothing more or less than communicating to it in a language that it and the human user can both "understand." And learning languages is one of the things children do best. Every normal child learns to talk. Why then should a child not learn to "talk" to a computer?

There are many reasons why someone might expect it to be difficult. For example, although babies learn to speak their native language with spectacular ease, most children have great difficulty learning foreign languages in schools and, indeed, often learn the written version of their own language none too successfully. Isn't learning a computer language more like the difficult process of learning a foreign written language than the easy one of learning to speak one's own language? And isn't the problem further compounded by all the difficulties most people encounter learning mathematics?

Two fundamental ideas run through this book. The first is that it is possible to design computers so that learning to communicate with them can be a natural process, more like learning French by living in France than like trying to learn it through the unnatural process of American foreign-language instruction in classrooms. Second, learning to communicate with a computer may change the way other learning takes place. The computer can be a mathematics-speaking and an alphabetic-speaking entity. We are learning how to make computers with which children love to communicate. When this communication occurs, children learn mathematics as a living language. Moreover, mathematical communication and alphabetic communication are thereby both transformed from the alien and therefore difficult things they are for most children into natural and therefore easy ones. The idea of "talking mathematics" to a computer can be generalized to a view of learning mathematics in "Mathland"; that is to say, in a context which is to learning mathematics what living in France is to learning French.

In this book the Mathland metaphor will be used to question deeply engrained assumptions about human abilities. It is generally assumed that children cannot learn formal geometry until well into their school years and that most cannot learn it too well even then. But we can quickly see that these assumptions are based on extremely weak evidence by asking analogous questions about the ability of children to learn French. If we had to base our opinions on observation of how poorly children learned French in American schools, we would have to conclude that most people were incapable of mastering it. But we know that all normal children would learn it very easily if they lived in France. My conjecture is that much of what we now see as too "formal" or "too mathematical" will be learned just as easily when children grow up in the computer-rich world of the very near future.

I use the examination of our relationship with mathematics as a thematic example of how technological and social processes interact in the construction of ideas about human capacities. And mathematical examples will also help to describe a theory of how learning works and of how it goes wrong.

I take from Jean Piaget[1] a model of children as builders of their own intellectual structures. Children seem to be innately gifted learners, acquiring long before they go to school a vast quantity of knowledge by a process I call "Piagetian learning," or "learning without being taught." For example, children learn to speak, learn the intuitive geometry needed to get around in space, and learn enough of logic and rhetorics to get around parents—all this without being "taught." We must ask why some learning takes place so early and spontaneously while some is delayed many years or does not happen at all without deliberately imposed formal instruction.

If we really look at the "child as builder" we are on our way to an answer. All builders need materials to build with. Where I am at variance with Piaget is in the role I attribute to the surrounding cultures as a source of these materials. In some cases the culture supplies them in abundance, thus facilitating constructive Piagetian learning. For example, the fact that so many important things (knives and forks, mothers and fathers, shoes and socks) come in pairs is a "material" for the construction of an intuitive sense of number. But in many cases where Piaget would explain the slower development of a particular concept by its greater complexity or formality, I see the critical factor as the relative poverty of the culture in those materials that would make the concept simple and concrete. In yet other cases the culture may provide materials but block their use. In the case of formal mathematics, there is both a shortage of formal materials and a cultural block as well. The mathophobia endemic in contemporary culture blocks many people from learning anything they recognize as "math," although they may have no trouble with mathematical knowledge they do not perceive as such.

We shall see again and again that the consequences of mathophobia go far beyond obstructing the learning of mathematics and science. They interact with other endemic "cultural toxins," for example, with popular theories of aptitudes, to contaminate peoples' images of themselves as learners. Difficulty with school math is often the first step of an invasive intellectual process that leads us all to define ourselves as bundles of aptitudes and ineptitudes, as being "mathematical" or "not mathematical," "artistic" or "not artistic," "musical" or "not musical," "profound" or "superficial," "intelligent" or "dumb." Thus deficiency becomes identity and learning is transformed from the early child's free exploration of the world to a chore beset by insecurities and self-imposed restrictions.

Two major themes—that children can learn to use computers in a masterful way, and that learning to use computers can change the way they learn everything else—have shaped my research agenda on computers and education. Over the past ten years I have had the good fortune to work with a group of colleagues and students at MIT (the LOGO[2] group in the Artificial Intelligence Laboratory) to create environments in which children can learn to communicate with computers. The metaphor of imitating the way the child learns to talk has been constantly with us in this work and has led to a vision of education and of education research very different from the traditional ones. For people in the teaching professions, the word "education" tends to evoke "teaching," particularly classroom teaching. The goal of education research tends therefore to be focused on how to improve classroom teaching. But if, as I have stressed here, the model of successful learning is the way a child learns to talk, a process that takes place without deliberate and organized teaching, the goal set is very different. I see the classroom as an artificial and inefficient learning environment that society has been forced to invent because its informal environments fail in certain essential learning domains, such as writing or grammar or school math. I believe that the computer presence will enable us to so modify the learning environment outside the classrooms that much if not all the knowledge schools presently try to teach with such pain and expense and such limited success will be learned, as the child learns to talk, painlessly, successfully, and without organized instruction. This obviously implies that schools as we know them today will have no place in the future. But it is an open question whether they will adapt by transforming themselves into something new or wither away and be replaced.

Although technology will play an essential role in the realization of my vision of the future of education, my central focus is not on the machine but on the mind, and particularly on the way in which intellectual movements and cultures define themselves and grow. Indeed, the role I give to the computer is that of a *carrier* of cultural "germs" or "seeds" whose intellectual products will not need technological support once they take root in an actively growing mind. Many if not all the children who grow up with a love and aptitude for mathematics owe this feeling, at least in part, to the fact that they happened to acquire "germs" of the "math culture" from adults, who, one might say, knew how to speak mathematics, even if only in the way that Moliere had M. Jourdain speak prose without knowing it. These "math-speaking" adults do not necessarily know how to solve equations; rather, they are marked by a turn of mind that shows up in the logic of their arguments and in the fact that for them to play is often to play with such things as puzzles, puns, and paradoxes. Those children who prove recalcitrant to math and science education include many whose environments happened to be relatively poor in math-speaking adults. Such children come to school lacking elements necessary for the easy learning of school math. School has been unable to supply these missing elements, and, by forcing the children into learning situations doomed in advance, it generates powerful negative feelings about mathematics and perhaps about learning in general. Thus is set up a vicious self-perpetuating cycle. For these same children will one day be parents and will not only fail to pass on mathematical germs but will almost certainly infect their children with the opposing and intellectually destructive germs of mathophobia.

Fortunately it is sufficient to break the self-perpetuating cycle at one point for it to remain broken forever. I shall show how computers might enable us to do this, thereby breaking the vicious cycle without creating a dependence on machines. My discussion differs from most arguments about "nature versus nurture" in two ways. I shall be much more specific both about what kinds of nurturance are needed for intellectual growth and about what can be done to create such nurturance in the home as well as in the wider social context.

Thus this book is really about how a culture, a way of thinking, an idea comes to inhabit a young mind. I am suspicious of thinking about such problems too abstractly, and I shall write here with particular restricted focus. I shall in fact concentrate on those ways of thinking that I know best. I begin by looking at what I know about my own development. I do this in all humility, without any implication that what I have become is what everyone should become. But I think that the best way to understand learning is first to understand specific, well-chosen cases and then to worry afterward about how to generalize from this understanding. You can't think seriously about thinking without thinking about thinking about something. And the something I know best how to think about is mathematics. When in this book I write of mathematics, I do not think of myself as writing for an audience of mathematicians interested in mathematical thinking for its own sake. My interest is in universal issues of how people think and how they learn to think.

When I trace how I came to be a mathematician, I see much that was idiosyncratic, much that could not be duplicated as part of a generalized vision of education reform. And I certainly don't think that we would want everyone to become a mathematician. But I think that the kind of pleasure I take in mathematics should be part of a general vision of what education should be about. If we can grasp the essence of one person's experiences, we may be able to replicate its consequences in other ways, and in particular this consequence of finding beauty in abstract things. And so I shall be writing quite a bit about mathematics. I give my apologies to readers who hate mathematics, but I couple that apology with an offer to help them learn to like it a little better—or at least to change their image of what "speaking mathematics" can be all about.

In the Foreword of this book I described how gears helped mathematical ideas to enter my life. Several qualities contributed to their effectiveness. First, they were part of my natural "landscape," embedded in the culture around me. This made it possible for me to find them myself and relate to them in my own fashion. Second, gears were part of the world of adults around me and through them I could relate to these people. Third, I could use my body to think about the gears. I could feel how gears turn by imagining my body turning. This made it possible for me to draw on my "body knowledge" to think about gear systems. And finally, because, in a very real sense, the relationship between gears contains a great deal of mathematical information, I could use the gears to think about formal systems. I have described the way in which the gears served as an "object-to-think-with." I made them that for myself in my own development as a mathematician. The gears have also served me as an object-to-think-with in my work as an educational researcher. My goal has been the design of other objects that children can make theirs for themselves and in their own ways. Much of this book will describe my path through this kind of research. I begin by describing one example of a constructed computational "object-to-think-with." This is the "Turtle."[3]

The central role of the Turtle in this book should not be taken to mean that I propose it as a panacea for all educational problems. I see it as a valuable educational object, but its principal role here is to serve as a model for other objects, yet to be invented. My interest is in the process of invention of "objects-to-think-with," objects in which there is an intersection of cultural presence, embedded knowledge, and the possibility for personal identification.

The Turtle is a computer-controlled cybernetic animal. It exists within the cognitive minicultures of the "LOGO environment," LOGO being the computer language in which communication with the Turtle takes place. The Turtle serves no other purpose than of being good to program and good to think with. Some Turtles are abstract objects that live on computer screens. Others, like the floor Turtles shown in the frontispiece are physical objects that can be picked up like any mechanical toy. A first encounter often begins by showing the child how a Turtle can be made to move by typing commands at a keyboard. FORWARD 100 makes the Turtle move in a straight line a distance of 100 Turtle steps of about a millimeter each. Typing RIGHT 90 causes the Turtle to pivot in place through 90 degrees. Typing PENDOWN causes the Turtle to lower a pen so as to leave a visible trace of its path while PENUP instructs it to raise the pen. Of course the child needs to explore a great deal before gaining mastery of what the numbers mean. But the task is engaging enough to carry most children through this learning process.

The idea of programming is introduced through the metaphor of teaching the Turtle a new word. This is simply done, and children often begin their programming experience by programming the Turtle to respond to new commands invented by the child such as SQUARE or TRIANGLE or SQ or TRI or whatever the child wishes, by drawing the appropriate shapes. New commands once defined can be used to define others. For example just as the house in Figure 1 is built out of a triangle and a square, the program for drawing it is built out of the commands for drawing a square and a triangle. Figure 1 shows four steps in the evolution of this program. From these simple drawings the young programmer can go on in many different directions. Some work on more complex drawings, either figural or abstract. Some abandon the use of the Turtle as a drawing instrument and learn to use its touch sensors to program it to seek out or avoid objects.[4] Later children learn that the computer can be programmed to make music as well as move Turtles and combine the two activities by programming Turtles to dance. Or they can move on from floor Turtles to "screen Turtles," which they program to draw moving pictures in bright colors. The examples are infinitely varied, but in each the child is learning how to exercise control over an exceptionally rich and sophisticated "micro-world."

Readers who have never seen an interactive computer display might find it hard to imagine where this can lead. As a mental exercise they might like to imagine an electronic sketchpad, a computer graphics display of the not-too-distant future. This is a television screen that can display moving pictures in color. You can also "draw" on it, giving it instructions, perhaps by typing, perhaps by speaking, or perhaps by pointing with a wand. On request, a palette of colors could appear on the screen. You can choose a color by pointing at it with the wand. Until you change your choice, the wand draws in that color. Up to this point the distinction from traditional art materials may seem slight, but the distinction becomes very real when you begin to think about editing the drawing. You can "talk to your drawing" in computer language. You can "tell" it to replace this color with that. Or set a drawing in motion. Or make two copies and set them in counterrotating motion. Or replace the color palette with a sound palette and "draw" a piece of music. You can file your work in computer memory and retrieve it at your pleasure, or have it delivered into the memory of any of the many millions of other computers linked to the central communication network for the pleasure of your friends.

That all this would be fun needs no argument. But it is more than fun. Very powerful kinds of learning are taking place. Children working with an electronic sketchpad are learning a language for talking about shapes and fluxes of shapes, about velocities and rates of change, about processes and procedures. They are learning to speak mathematics, and acquiring a new image of themselves as mathematicians.

In my description of children working with Turtles, I implied that children can learn to program. For some readers this might be tantamount to the suspension of disbelief required when we enter a theater to watch a play. For them programming is a complex and marketable skill acquired by some mathematically gifted adults. But my experience is very different. I have seen hundreds of elementary school children learn very easily to program, and evidence is accumulating to indicate that much younger children could do so as well. The children in these studies are not exceptional, or rather, they are exceptional in every conceivable way. Some of the children were highly successful in school, some were diagnosed as emotionally or cognitively disabled. Some of the children were so severely afflicted by cerebral palsy that they had never purposefully manipulated physical objects. Some of them had expressed their talents in "mathematical" forms, some in "verbal" forms, and some in artistically "visual" or in "musical" forms.

Figure 1

A Plan

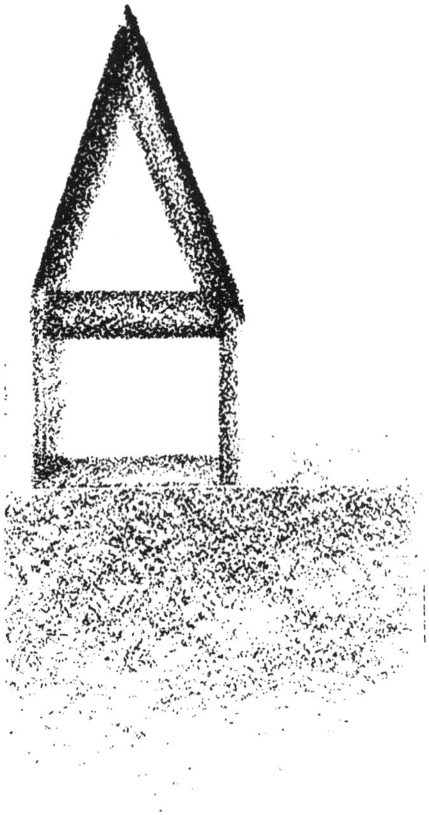

A Bug

TO HOUSE
SQ
TRI

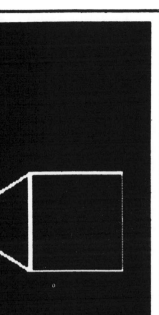

```
TO HOUSE
SQ
RIGHT 30
TR1
END
```

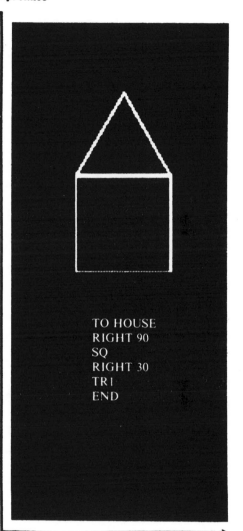

```
TO HOUSE
RIGHT 90
SQ
RIGHT 30
TR1
END
```

Of course these children did not achieve a fluency in programming that came close to matching their use of spoken language. If we take the Mathland metaphor seriously, their computer experience was more like learning French by spending a week or two on vacation in France than like living there. But like children who have spent a vacation with foreign-speaking cousins, they were clearly on their way to "speaking computer."

When I have thought about what these studies mean I am left with two clear impressions. First, that all children will, under the right conditions, acquire a proficiency with programming that will make it one of their more advanced intellectual accomplishments. Second, that the "right conditions" are very different from the kind of access to computers that is now becoming established as the norm in schools. The conditions necessary for the kind of relationships with a computer that I will be writing about in this book require more and freer access to the computer than educational planners currently anticipate. And they require a kind of computer language and a learning environment around that language very different from those the schools are now providing. They even require a kind of computer rather different from those that the schools are currently buying.

It will take most of this book for me to convey some sense of the choices among computers, computer languages, and more generally, among computer cultures, that influence how well children will learn from working with computation and what benefits they will get from doing so. But the question of the *economic* feasibility of free access to computers for every child can be dealt with immediately. In doing so I hope to remove any doubts readers may have about the "economic realism" of the "vision of education" I have been talking about.

My vision of a new kind of learning environment demands free contact between children and computers. This could happen because the child's family buys one or a child's friends have one. For purposes of discussion here (and to extend our discussion to all social groups) let us assume that it happens because schools give every one of their students his or her own powerful personal computer. Most "practical" people (including parents, teachers, school principals, and foundation administrators) react to this idea in much the same way: "Even if computers could have all the effects you talk about, it would still be impossible to put your ideas into action. Where would the money come from?"

What these people are saying needs to be faced squarely. They are wrong. Let's consider the cohort of children who will enter kindergarten in the year 1987, the "Class of 2000," and let's do some arithmetic. The direct public cost of schooling a child for thirteen years, from kindergarten through twelfth grade is over $20,000 today (and for the class of 2000, it may be closer to $30,000). A conservatively high estimate of the cost of supplying each of these children with a personal computer with enough power for it to serve the kinds of educational ends described in this book, and of upgrading, repairing, and replacing it when necessary would be about $1,000 per student, distributed over thirteen years in school. Thus, "computer costs" for the class of 2,000 would represent only about 5 percent of the total public expenditure on education, and this would be the case even if nothing else in the structure of educational costs changed because of the computer presence. But in fact computers in education stand a good chance of making other aspects of education cheaper. Schools might be able to reduce their cycle from thirteen years to twelve years; they might be able to take advantage of the greater autonomy the computer gives students and increase the size of classes by one or two students without decreasing the personal attention each student is given. Either of these two moves would "recuperate" the computer cost.

My goal is not educational economies: It is not to use computation to shave a year off the time a child spends in an otherwise unchanged school or to push an extra child into an elementary school classroom. The point of this little exercise in educational "budget balancing" is to do something to the state of mind of my readers as they turn to the first chapter of this book. I have described myself as an educational utopian—not because I have projected a future of education in which children are surrounded by high technology, but because I believe that certain uses of very powerful computational technology and computational ideas can provide children with new possibilities for learning, thinking, and growing emotionally as well as cognitively. In the chapters that follow I shall try to give you some idea of these possibilities, many of which are dependent on a computer-rich future, a future where a computer will be a significant part of every child's life. But I want my readers to be very clear that what is "utopian" in my vision and in this book is a particular way of using computers, of forging new relationships between computers and people—that the computer will be there to be used is simply a conservative premise.

Chapter 1

Computers and Computer Cultures

IN MOST contemporary educational situations where children come into contact with computers the computer is used to put children through their paces, to provide exercises of an appropriate level of difficulty, to provide feedback, and to dispense information. The computer programming the child. In the LOGO environment the relationship is reversed: The child, even at preschool ages, is in control: The child programs the computer. And in teaching the computer how to think, children embark on an exploration about how they themselves think. The experience can be heady: Thinking about thinking turns the child into an epistemologist, an experience not even shared by most adults.

This powerful image of child as epistemologist caught my imagination while I was working with Piaget. In 1964, after five years at Piaget's Center for Genetic Epistemology in Geneva, I came away impressed by his way of looking at children as the active builders of their own intellectual structures. But to say that intellectual structures are built by the learner rather than taught by a teacher does not mean that they are built from nothing. On the contrary: Like other builders, children appropriate to their own use materials they find about them, most saliently the models and metaphors suggested by the surrounding culture.

Piaget writes about the order in which the child develops different intellectual abilities. I give more weight than he does to the influence of the materials a particular culture provides in determining that order. For example, our culture is very rich in materials useful for the child's construction of certain components of numerical and logical thinking. Children learn to count; they learn that the result of counting is independent of order and special arrangement; they extend this "conservation" to thinking about the properties of liquids as they are poured and of solids which change their shape. Children develop these components of thinking preconsciously and "spontaneously," that is to say without deliberate teaching. Other components of knowledge, such as the skills involved in doing permutations and combinations, develop more slowly, or do not develop at all without formal schooling. Taken as a whole this book is an argument that in many important cases this developmental difference can be attributed to our culture's relative poverty in materials from which the apparently "more advanced" intellectual structures can be built. This argument will be very different from cultural interpretations of Piaget that look for differences between city children in Europe or the United States and tribal children in African jungles. When I speak here of "our" culture I mean something less parochial. I am not trying to contrast New York with Chad. I am interested in the difference between precomputer cultures (whether in American cities or African tribes) and the "computer cultures" that may develop everywhere in the next decades.

I have already indicated one reason for my belief that the computer presence might have more fundamental effects on intellectual development than did other new technologies, including television and even printing. The metaphor of computer as mathematics-speaking entity puts the learner in a qualitatively new kind of relationship to an important domain of knowledge. Even the best of educational television is limited to offering quantitative improvements in the kinds of learning that existed without it. "Sesame Street" might offer better and more engaging explanations than a child can get from some parents or nursery school teachers, but the child is still in the position of listening to explanations. By contrast, when a child learns to program, the process of learning is transformed. It becomes more active and self-directed. In particular, the knowledge is acquired for a recognizable personal purpose. The child does something with it. The new knowledge is a source of power and is experienced as such from the moment it begins to form in the child's mind.

I have spoken of mathematics being learned in a new way. But much more is affected than mathematics. One can get an idea of the extent of what is changed by examining another of Piaget's ideas. Piaget distinguishes between "concrete" thinking and "formal" thinking. Concrete thinking is already well on its way by the time the child enters the first grade at age 6 and is consolidated in the following several years. Formal thinking does not develop until the child is almost twice as old, that is to say at age 12, give or take a year or two, and some researchers have even suggested that many people never achieve fully formal thinking. I do not fully accept Piaget's distinction, but I am sure that it is close enough to reality to help us make sense of the idea that the consequences for intellectual development of one innovation could be qualitatively greater than the cumulative quantitative effects of a thousand others. Stated most simply, my conjecture is that the computer can concretize (and personalize) the formal. Seen in this light, it is not just another powerful educational tool. It is unique in providing us with the means for addressing what Piaget and many others see as the obstacle which is overcome in the passage from child to adult thinking. I believe that it can allow us to shift the boundary separating concrete and formal. Knowledge that was accessible only through formal processes can now be approached concretely. And the real magic comes from the fact that this knowledge includes those elements one needs to become a formal thinker.

This description of the role of the computer is rather abstract. I shall concretize it, anticipating discussions which occur in later chapters of this book, by looking at the effect of working with computers on two kinds of thinking Piaget associates with the formal stage of intellectual development: combinatorial thinking, where one has to reason in terms of the set of all possible states of a system, and self-referential thinking about thinking itself.

In a typical experiment in combinatorial thinking, children are asked to form all the possible combinations (or "families") of beads of assorted colors. It really is quite remarkable that most children are unable to do this systematically and accurately until they are in the fifth or sixth grades. Why should this be? Why does this task seem to be so much more difficult than the intellectual feats accomplished by seven and eight year old children? Is its logical structure essentially more complex? Can it possibly require a neurological mechanism that does not mature until the approach of puberty? I think that a more likely explanation is provided by looking at the nature of the culture. The task of making the families of beads can be looked at as constructing and executing a program, a very common sort of program, in which two loops are nested: Fix a first color and run through all the possible second colors, then repeat until all possible first colors have been run through. For someone who is thoroughly used to computers and programming there is nothing "formal" or abstract about this task. For a child in a computer culture it would be as concrete as matching up knives and forks at the dinner table. Even the common "bug" of including some families twice (for example, red-blue and blue-red) would be well-known. Our culture is rich in pairs, couples, and one-to-one correspondences of all sorts, and it is rich in language for talking about such things. This richness provides both the incentive and a supply of models and tools for children to build ways to think about such issues as whether three large pieces of candy are more or less than four much smaller pieces. For such problems our children acquire an excellent intuitive sense of quantity. But our culture is relatively poor in models of systematic procedures. Until recently there was not even a name in popular language for programming, let alone for the ideas needed to do so successfully. There is no word for "nested loops" and no word for the double-counting bug. Indeed, there are no words for the powerful ideas computerists refer to as "bug" and "debugging."

Without the incentive or the materials to build powerful, concrete ways to think about problems involving systematicity, children are forced to approach such problems in a groping, abstract fashion. Thus cultural factors that are common to both the American city and the African village can explain the difference in age at which children build their intuitive knowledge of quantity and of systematicity.

While still working in Geneva I had become sensitive to the way in which materials from the then very young computer cultures were allowing psychologists to develop new ways to think about thinking.[1] In fact, my entry into the world of computers was motivated largely by the idea that children could also benefit, perhaps even more than the psychologists, from the way in which computer models seemed able to give concrete form to areas of knowledge that had previously appeared so intangible and abstract.

I began to see how children who had learned to program computers could use very concrete computer models to think about thinking and to learn about learning and in doing so, enhance their powers as psychologists and as epistemologists. For example, many children are held back in their learning because they have a model of learning in which you have either "got it" or "got it wrong." But when you learn to program a computer you almost never get it right the first time. Learning to be a master programmer is learning to become highly skilled at isolating and correcting "bugs," the parts that keep the program from working. The question to ask about the program is not whether it is right or wrong, but if it is fixable. If this way of looking at intellectual products were generalized to how the larger culture thinks about knowledge and its acquisition, we all might be less intimidated by our fears of "being wrong." This potential influence of the computer on changing our notion of a black and white version of our successes and failures is an example of using the computer as an "object-to-think-with." It is obviously not necessary to work with computers in order to acquire good strategies for learning. Surely "debugging" strategies were developed by successful learners long before computers existed. But thinking about learning by analogy with developing a program is a powerful and accessible way to get started on becoming more articulate about one's debugging strategies and more deliberate about improving them.

My discussion of a computer culture and its impact on thinking presupposes a massive penetration of powerful computers into people's lives. That this will happen there can be no doubt. The calculator, the electronic game, and the digital watch were brought to us by a technical revolution that rapidly lowered prices for electronics in a period when all others were rising with inflation. That same technological revolution, brought about by the integrated circuit, is now bringing us the personal computer. Large computers used to cost millions of dollars because they were assembled out of millions of physically distinct parts. In the new technology a complex circuit is not assembled but made as a whole, solid entity—hence the term "integrated circuit." The effect of integrated circuit technology on cost can be understood by comparing it to printing. The main expenditure in making a book occurs long before the press begins to roll. It goes into writing, editing, and typesetting. Other costs occur after the printing: binding, distributing, and marketing. The actual cost per copy for printing itself is negligible. And the same is true for a powerful as for a trivial book. So, too, most of the cost of an integrated circuit goes into a preparatory process; the actual cost of making an individual circuit becomes negligible, provided enough are sold to spread the costs of development. The consequences of this technology for the cost of computation are dramatic. Computers that would have cost hundreds of thousands in the 1960s and tens of thousands in the early 1970s can now be made for less than a dollar. The only limiting factor is whether the particular circuit can fit onto what corresponds to a "page"—that is to say the "silicon chips" on which the circuits are etched.

But each year in a regular and predictable fashion the art of etching circuits on silicon chips is becoming more refined. More and more complex circuitry can be squeezed onto a chip, and the computer power that can be produced for less than a dollar increases. I predict that long before the end of the century, people will buy children toys with as much computer power as the great IBM computers currently selling for millions of dollars. And as for computers to be used as such, the main cost of these machines will be the peripheral devices, such as the keyboard. Even if these do not fall in price, it is likely that a supercomputer will be equivalent in price to a typewriter and a television set.

There really is no disagreement among experts that the cost of computers will fall to a level where they will enter everyday life in vast numbers. Some will be there as computers proper, that is to say, programmable machines. Others might appear as games of ever-increasing complexity and in automated supermarkets where the shelves, maybe even the cans, will talk. One really can afford to let one's imagination run wild. There is no doubt that the material surface of life will become very different for everyone, perhaps most of all for children. But there has been significant difference of opinion about the effects this computer presence will produce. I would distinguish my thinking from two trends of thinking which I refer to here as the "skeptical" and the "critical."

Skeptics do not expect the computer presence to make much difference in how people learn and think. I have formulated a number of possible explanations for why they think as they do. In some cases I think the skeptics might conceive of education and the effect of computers on it too narrowly. Instead of considering general cultural effects, they focus attention on the use of the computer as a device for programmed instruction. Skeptics then conclude that while the computer might produce some improvements in school learning, it is not likely to lead to fundamental change. In a sense, too, I think the skeptical view derives from a failure to appreciate just how much Piagetian learning takes place as a child grows up. If a person conceives of children's intellectual development (or, for that matter, moral or social development) as deriving chiefly from deliberate teaching, then such a person would be likely to underestimate the potential effect that a massive presence of computers and other interactive objects might have on children.

The critics,[2] on the other hand, do think that the computer presence will make a difference and are apprehensive. For example, they fear that more communication via computers might lead to less human association and result in social fragmentation. As knowing how to use a computer becomes increasingly necessary to effective social and economic participation, the position of the underprivileged could worsen, and the computer could exacerbate existing class distinctions. As to the political effect computers will have, the critics' concerns resonate with Orwellian images of a 1984 where home computers will form part of a complex system of

25

surveillance and thought control. Critics also draw attention to potential mental health hazards of computer penetration. Some of these hazards are magnified forms of problems already worrying many observers of contemporary life; others are problems of an essentially new kind. A typical example of the former kind is that our grave ignorance of the psychological impact of television becomes even more serious when we contemplate an epoch of super TV. The holding power and the psychological impact of the television show could be increased by the computer in at least two ways. The content might be varied to suit the tastes of each individual viewer, and the show might become interactive, drawing the "viewer" into the action. Such things belong to the future, but people who are worried about the impact of the computer on people already cite cases of students spending sleepless nights riveted to the computer terminal, coming to neglect both studies and social contact. Some parents have been reminded of these stories when they observe a special quality of fascination in their own children's reaction to playing with the still rudimentary electronic games.

In the category of problems that are new rather than aggravated versions of old ones, critics have pointed to the influence of the allegedly mechanized thought processes of computers on how people think. Marshall McCluhan's dictum that "the medium is the message" might apply here: If the medium is an interactive system that takes in words and speaks back like a person, it is easy to get the message that machines are like people and that people are like machines. What this might do to the development of values and self-image in growing children is hard to assess. But it is not hard to see reasons for worry.

Despite these concerns I am essentially optimistic—some might say utopian—about the effect of computers on society. I do not dismiss the arguments of the critics. On the contrary, I too see the computer presence as a potent influence on the human mind. I am very much aware of the holding power of an interactive computer and of how taking the computer as a model can influence the way we think about ourselves. In fact the work on LOGO to which I have devoted much of the past ten years consists precisely of developing such forces in positive directions. For example, the critic is horrified at the thought of a child hypnotically held by a futuristic, computerized super-pinball machine. In the LOGO work we have invented versions of such machines in which powerful ideas from physics or mathematics or linguistics are embedded in a way that permits the player to learn them in a natural fashion, analogous to how a child learns to speak. The computer's "holding power," so feared by critics, becomes a useful educational tool. Or take another, more profound example. The critic is afraid that children will adopt the computer as model and eventually come to "think mechanically" themselves. Following the opposite tack, I have invented ways to take educational advantage of the opportunities to master the art of *deliberately* thinking like a computer, according, for example, to the stereotype of a computer program that proceeds in a step-by-step, literal, mechanical fashion. There are situations where this style of thinking is appropriate and useful. Some children's difficulties in learning formal subjects such as grammar or mathematics derive from their inability to see the point of such a style.

A second educational advantage is indirect but ultimately more important. By deliberately learning to imitate mechanical thinking, the learner becomes able to articulate what mechanical thinking is and what it is not. The exercise can lead to greater confidence about the ability to choose a cognitive style that suits the problem. Analysis of "mechanical thinking" and how it is different from other kinds and practice with problem analysis can result in a new degree of intellectual sophistication. By providing a very concrete, down-to-earth model of a particular style of thinking, work with the computer can make it easier to understand that there is such a thing as a "style of thinking." And giving children the opportunity to choose one style or another provides an opportunity to develop the skill necessary to choose between styles. Thus instead of inducing mechanical thinking, contact with computers could turn out to be the best conceivable antidote to it. And for me what is most important in this is that through these experiences these children would be serving their apprenticeships as epistemologists, that is to say learning to think articulately about thinking.

The intellectual environments offered to children by today's cultures are poor in opportunities to bring their thinking about thinking into the open, to learn to talk about it and to test their ideas by externalizing them. Access to computers can dramatically change this situation. Even the simplest Turtle work can open new opportunities for sharpening one's thinking about thinking: Programming the Turtle starts by making one reflect on how one does oneself what one would like the Turtle to do. Thus teaching the Turtle to act or to "think" can lead one to reflect on one's own actions and thinking. And as children move on, they program the computer to make more complex decisions and find themselves engaged in reflecting on more complex aspects of their own thinking.

In short, while the critic and I share the belief that working with computers can have a powerful influence on how people think, I have turned my attention to exploring how this influence could be turned in positive directions.

I see two kinds of counterarguments to my arguments against the critics. The first kind challenges my belief that it is a good thing for children to be epistemologists. Many people will argue that overly analytic, verbalized thinking is counterproductive even if it is deliberately chosen. The second kind of objection challenges my suggestion that computers are likely to lead to more reflective self-conscious thinking. Many people will argue that work with computers usually has the opposite effect. These two kinds of objections call for different kinds of analysis and cannot be discussed simultaneously. The first kind raises technical questions about the psychology of learning which will be discussed in chapters 4 and 6. The second kind of objection is most directly answered by saying that there is absolutely no inevitability that computers will have the effects I hope to see. Not all computer systems do. Most in use today do not. In LOGO environments I have seen children engaged in animated conversations about their own personal knowledge as they try to capture it in a program to make a Turtle carry out an action that they themselves know very well how to do. But of course the physical presence of a computer is not enough to insure that such conversations will come about. Far from it. In thousands of schools and in tens of thousands of private homes children are right now living through very different computer experiences. In most cases the computer is being used either as a versatile video game or as a "teaching machine" programmed to put children through their paces in arithmetic or spelling. And even when children are taught by a parent, a peer, or a professional teacher to write simple programs in a language like BASIC, this activity is not accompanied at all by the kind of epistemological reflection that we see in the LOGO environments. So I share a skepticism with the critics about what is being done with computation now. But I am interested in stimulating a major change in how things can be. The bottom line for such changes is political. What is happening now is an empirical question. What can happen is a technical question. But what will happen is a political question, depending on social choices.

The central open questions about the effect of computers on children in the 1980s are these: Which people will be attracted to the world of computers, what talents will they bring, and what tastes and ideologies will they impose on the growing computer culture? I have described children in LOGO environments engaged in self-referential discussions about their own thinking. This could happen because the LOGO language and the Turtle were designed by people who enjoy such discussion and worked hard to design a medium that would encourage it. Other designers of computer systems have different tastes and different ideas about what kinds of activities are suitable for children. Which design will prevail, and in what sub-culture, will not be decided by a simple bureaucratic decision made, for example, in a government Department of Education or by a committee of experts. Trends in computer style will emerge from a complex web of decisions by Foundations with resources to support one or another design, by corporations who may see a market, by schools, by individuals who will decide to make their career in the new field of activity, and by children who will have their own say in what they pick up and what they make of it. People often ask whether in the future children will program computers or become absorbed in pre-programmed activities. The answer must be that some children will do the one, some the other, some both and some neither. But which children, and most importantly, which social classes of children, will fall into each category will be influenced by the kind of computer activities and the kind of environments created around them.

As an example, we consider an activity which may not occur to most people when they think of computers and children: the use of a computer as a writing instrument. For me, writing means making a rough draft and refining it over a considerable period of time. My image of myself as a writer includes the expectation of an "unacceptable" first draft that will develop with successive editing into presentable form. But I would not be able to afford this image if I were a third grader. The physical act of writing would be slow and laborious. I would have no secretary. For most children rewriting a text is so laborious that the first draft is the final copy, and the skill of rereading with a critical eye is never acquired. This changes dramatically when children have access to computers capable of manipulating text. The first draft is composed at the keyboard. Corrections are made easily. The current copy is always neat and tidy. I have seen a child move from total rejection of writing to an intense involvement (accompanied by rapid improvement of quality) within a few weeks of beginning to write with a computer. Even more dramatic changes are seen when the child has physical handicaps that make writing by hand more than usually difficult or even impossible.

This use of computers is rapidly becoming adopted wherever adults write for a living. Most newspapers now provide their staff with "word processing" computer systems. Many writers who work at home are acquiring their own computers, and the computer terminal is steadily displacing the typewriter as the secretary's basic tool. The image of children using the computer as a writing instrument is a particularly good example of my general thesis that what is good for professionals is good for children. But this image of how the computer might contribute to children's mastery of language is dramatically opposed to the one that is taking root in most elementary schools. There the computer is seen as a teaching instrument. It gives children practice in distinguishing between verbs and nouns, in spelling, and in answering multiple-choice questions about the meaning of pieces of text. As I see it, this difference is not a matter of a small and technical choice between two teaching strategies. It reflects a fundamental difference in educational philosophies. More to the point, it reflects a difference in views on the nature of childhood. I believe that the computer as writing instrument offers children an opportunity to become more like adults, indeed like advanced professionals, in their relationship to their intellectual products and to themselves. In doing so, it comes into head-on collision with the many aspects of school whose effect, if not whose intention, is to "infantilize" the child.

Word processors *can* make a child's experience of writing more like that of a real writer. But this can be undermined if the adults surrounding that child fail to appreciate what it is like to be a writer. For example, it is only too easy to imagine adults, including teachers, expressing the view that editing and re-editing a text is a waste of time ("Why don't you get on to something new?" or "You aren't making it any better, why don't you fix your spelling?").

As with writing, so with music-making, games of skill, complex graphics, whatever: The computer is not a culture unto itself but it can serve to advance very different cultural and philosophical outlooks. For example, one could think of the Turtle as a device to teach elements of the traditional curriculum, such as notions of angle, shape, and coordinate systems. And in fact, most teachers who consult me about its use are, quite understandably, trying to use it in this way. Their questions are about classroom organization, scheduling problems, pedagogical issues raised by the Turtle's introduction, and especially, about how it relates conceptually to the rest of the curriculum. Of course the Turtle can help in the teaching of traditional curriculum, but I have thought of it as a vehicle for Piagetian learning, which to me is learning without curriculum.

There are those who think about creating a "Piagetian curriculum" or "Piagetian teaching methods." But to my mind these phrases and the activities they represent are contradictions in terms. I see Piaget as the theorist of learning without curriculum and the theorist of the kind of learning that happens without deliberate teaching. To turn him into the theorist of a new curriculum is to stand him on his head.

But "teaching without curriculum" does not mean spontaneous, free-form classrooms or simply "leaving the child alone." It means supporting children as they build their own intellectual structures with materials drawn from the surrounding culture. In this model, educational intervention means changing the culture, planting new constructive elements in it and eliminating noxious ones. This is a more ambitious undertaking than introducing a curriculum change, but one which is feasible under conditions now emerging.

Suppose that thirty years ago an educator had decided that the way to solve the problem of mathematics education was to arrange for a significant fraction of the population to become fluent in (and enthusiastic about) a new mathematical language. The idea might have been good in principle, but in practice it would have been absurd. No one had the power to implement it. Now things are different. Many millions of people are learning programming languages for reasons that have nothing to do with the education of children. Therefore, it becomes a practical proposition to influence the form of the languages they learn and the likelihood that their children will pick up these languages.

The educator must be an anthropologist. The educator as anthropologist must work to understand which cultural materials are relevant to intellectual development. Then, he or she needs to understand which trends are taking place in the culture. Meaningful intervention must take the form of working with these trends. In my role of educator as anthropologist I see new needs being generated by the penetration of the computer into personal lives. People who have computers at home or who use them at work will want to be able to talk about them to their children. They will want to be able to teach their children to use the machines. Thus there could be a cultural demand for something like Turtle graphics in a way there never was, and perhaps never could be, a cultural demand for the New Math.

Throughout the course of this chapter I have been talking about the ways in which choices made by educators, foundations, governments, and private individuals can affect the potentially revolutionary changes in how children learn. But making good choices is not always easy, in part because past choices can often haunt us. There is a tendency for the first usable, but still primitive, product of a new technology to dig itself in. I have called this phenomenon the QWERTY phenomenon.

The top row of alphabetic keys of the standard typewriter reads QWERTY. For me this symbolizes the way in which technology can all too often serve not as a force for progress but for keeping things stuck. The QWERTY arrangement has no rational explanation, only a historical one. It was introduced in response to a problem in the early days of the typewriter: The keys used to jam. The idea was to minimize the collision problem by separating those keys that followed one another frequently. Just a few years later, general improvements in the technology removed the jamming problem, but QWERTY stuck. Once adopted, it resulted in many millions of typewriters and a method (indeed a full-blown curriculum) for learning typing. The social cost of change (for example, putting the most used keys *together* on the keyboard) mounted with the vested interest created by the fact that so many fingers now knew how to follow the QWERTY keyboard. QWERTY has stayed on despite the existence of other, more "rational" systems. On the other hand, if you talk to people about the QWERTY arrangement they will justify it by "objective" criteria. They will tell you that it "optimizes this" or it "minimizes that." Although these justifications have no rational foundation, they illustrate a process, a social process, of myth construction that allows us to build a justification for primitivity into any system. And I think that we are well on the road to doing exactly the same thing with the computer. We are in the process of digging ourselves into an anachronism by preserving practices that have no rational basis beyond their historical roots in an earlier period of technological and theoretical development.

The use of computers for drill and practice is only one example of the QWERTY phenomenon in the computer domain. Another example occurs even when attempts are made to allow students to learn to program the computer. As we shall see in later chapters, learning to program a computer involves learning a "programming language." There are many such languages—for example, FORTRAN, PASCAL, BASIC, SMALLTALK, and LISP, and the lesser known language LOGO, which our group has used in most of our experiments with computers and children. A powerful QWERTY phenomenon is to be expected when we choose the language in which children are to learn to program computers. I shall argue in detail that the issue is consequential. A programming language is like a natural, human language in that it favors certain metaphors, images, and ways of thinking. The language used strongly colors the computer culture. It would seem to follow that educators interested in using computers and sensitive to cultural influences would pay particular attention to the choice of language. But nothing of the sort has happened. On the contrary, educators, too timid in technological matters or too ignorant to attempt to influence the languages offered by computer manufacturers, have accepted certain programming languages in much the same way as they accepted the QWERTY keyboard. An informative example is the way in which the programming language BASIC[3] has established itself as the obvious language to use in teaching American children how to program computers. The relevant technical information is this: A very small computer can be made to understand BASIC, while other languages demand more from the computer. Thus, in the early days when computer power was extremely expensive, there was a genuine technical reason for the use of BASIC, particularly in schools where budgets were always tight. Today, and in fact for several years now, the cost of computer memory has fallen to the point where any remaining economic advantages of using BASIC are insignificant. Yet in most high schools, the language remains almost synonymous with programming, despite the existence of other computer languages that are demonstrably easier to learn and are richer in the intellectual benefits that can come from learning them. The situation is paradoxical. The computer revolution has scarcely begun, but is already breeding its own conservatism. Looking more closely at BASIC provides a window on how a conservative social system appropriates and tries to neutralize a potentially revolutionary instrument.

BASIC is to computation what QWERTY is to typing. Many teachers have learned BASIC, many books have been written about it, many computers have been built in such a way that BASIC is "hardwired" into them. In the case of the typewriter, we noted how people invent "rationalizations" to justify the status quo. In the case of BASIC, the phenomenon has gone much further, to the point where it resembles ideology formation. Complex arguments are invented to justify features of BASIC that were originally included because the primitive technology demanded them or because alternatives were not well enough known at the time the language was designed.

An example of BASIC ideology is the argument that BASIC is easy to learn because it has a very small vocabulary. The surface validity of the argument is immediately called into question if we apply it to the context of how children learn natural languages. Imagine a suggestion that we invent a special language to help children learn to speak. This language would have a small vocabulary of just fifty words, but fifty words so well chosen that all ideas could be expressed using them. Would this language be easier to learn? Perhaps the vocabulary might be easy to learn, but the use of the vocabulary to express what one wanted to say would be so contorted that only the most motivated and brilliant children would learn to say more than "hi." This is close to the situation with BASIC. Its small vocabulary can be learned quickly enough. But using it is a different matter. Programs in BASIC acquire so labyrinthine a structure that in fact only the most motivated and brilliant ("mathematical") children do learn to use it for more than trivial ends.

One might ask why the teachers do not notice the difficulty children have in learning BASIC. The answer is simple: Most teachers do not expect high performance from most students, especially in a domain of work that appears to be as "mathematical" and "formal" as programming. Thus the culture's general perception of mathematics as inaccessible bolsters the maintenance of BASIC, which in turn confirms these perceptions. Moreover, the teachers are not the only people whose assumptions and prejudices feed into the circuit that perpetuates BASIC. There are also the computerists, the people in the computer world who make decisions about what languages their computers will speak. These people, generally engineers, find BASIC quite easy to learn, partly because they are accustomed to learning such very technical systems and partly because BASIC's sort of simplicity appeals to their system of values. Thus, a particular subculture, one dominated by computer engineers, is influencing the world of education to favor those school students who are most like that subculture. The process is tacit, unintentional: It has never been publicly articulated, let alone evaluated. In all of these ways, the social embedding of BASIC has far more serious consequences than the "digging in" of QWERTY.

There are many other ways in which the attributes of the subcultures involved with computers are being projected onto the world of education. For example, the idea of the computer as an instrument for drill and practice that appeals to teachers because it resembles traditional teaching methods also appeals to the engineers who design computer systems: Drill and practice applications are predictable, simple to describe, efficient in use of the machine's resources. So the best engineering talent goes into the development of computer systems that are biased to favor this kind of application. The bias operates subtly. The machine designers do not actually decide what will be done in the classrooms. That is done by teachers and occasionally even by carefully controlled comparative research experiments. But there is an irony in these controlled experiments. They are very good at telling whether the small effects seen in best scores are real or due to chance. But they have no way to measure the undoubtedly real (and probably more massive) effects of the biases built into the machines.

We have already noted that the conservative bias being built into the use of computers in education has also been built into other new technologies. The first use of the new technology is quite naturally to do in a slightly different way what had been done before without it. It took years before designers of automobiles accepted the idea that they were cars, not "horseless carriages," and the precursors of modern motion pictures were plays acted as if before a live audience but actually in front of a camera. A whole generation was needed for the new art of motion pictures to emerge as something quite different from a linear mix of theater plus photography. Most of what has been done up to now under the name of "educational technology" or "computers in education" is still at the stage of the linear mix of old instructional methods with new technologies. The topics I shall be discussing are some of the first probings toward a more organic interaction of fundamental educational principles and new methods for translating them into reality.

We are at a point in the history of education when radical

change is possible, and the possibility for that change is directly tied to the impact of the computer. Today what is offered in the education "market" is largely determined by what is acceptable to a sluggish and conservative system. But this is where the computer presence is in the process of creating an environment for change. Consider the conditions under which a new educational idea can be put into practice today and in the near future. Let us suppose that today I have an idea about how children could learn mathematics more effectively and more humanely. And let us suppose that I have been able to persuade a million people that the idea is a good one. For many products such a potential market would guarantee success. Yet in the world of education today this would have little clout: A million people across the nation would still mean a minority in every town's school system, so there might be no effective channel for the million voices to be expressed. Thus, not only do good educational ideas sit on the shelves, but the process of invention is itself stymied. This inhibition of invention in turn influences the selection of people who get involved in education. Very few with the imagination, creativity, and drive to make great new inventions enter the field. Most of those who do are soon driven out in frustration. Conservatism in the world of education has become a self-perpetuating *social* phenomenon.

Fortunately, there is a weak link in the vicious circle. Increasingly, the computers of the very near future will be the private property of individuals, and this will gradually return to the individual the power to determine patterns of education. Education will become more of a private act, and people with good ideas, different ideas, exciting ideas will no longer be faced with a dilemma where they either have to "sell" their ideas to a conservative bureaucracy or shelve them. They will be able to offer them in an open marketplace directly to consumers. There will be new opportunities for imagination and originality. There might be a renaissance of thinking about education.

Chapter 2

Mathophobia: The Fear of Learning

PLATO WROTE over his door, "Let only geometers enter." Times have changed. Most of those who now seek to enter Plato's intellectual world neither know mathematics nor sense the least contradiction in their disregard for his injunction. Our culture's schizophrenic split between "humanities" and "science" supports their sense of security. Plato was a philosopher, and philosophy belongs to the humanities as surely as mathematics belongs to the sciences.

This great divide is thoroughly built into our language, our worldview, our social organization, our educational system, and, most recently, even our theories of neurophysiology. It is self-perpetuating: The more the culture is divided, the more each side builds separation into its new growth.

I have already suggested that the computer may serve as a force to break down the line between the "two cultures." I know that the humanist may find it questionable that a "technology" could change his assumptions about what kind of knowledge is relevant to his or her perspective of understanding people. And to the scientist dilution of rigor by the encroachment of "wishy-washy" humanistic thinking can be no less threatening. Yet the computer presence might, I think, plant seeds that could grow into a less dissociated cultural epistemology.

The status of mathematics in contemporary culture is one of the most acute symptoms of its dissociation. The emergence of a "humanistic" mathematics, one that is not perceived as separated from the study of man and "the humanities," might well be the sign that a change is in sight. So in this book I try to show how the computer presence can bring children into a more humanistic as well as a more humane relationship with mathematics. In doing so I shall have to go beyond discussion of mathematics. I shall have to develop a new perspective on the process of learning itself.

It is not uncommon for intelligent adults to turn into passive observers of their own incompetence in anything but the most rudimentary mathematics. Individuals may see the direct consequences of this intellectual paralysis in terms of limiting job possibilities. But the indirect, secondary consequences are even more serious. One of the main lessons learned by most people in math class is a sense of having rigid limitations. They learn a balkanized image of human knowledge which they come to see as a patchwork of territories separated by impassable iron curtains. My challenge is not to the sovereignty of the intellectual territories but to the restrictions imposed on easy movement among them. I do not wish to reduce mathematics to literature or literature to mathematics. But I do want to argue that their respective ways of thinking are not as separate as is usually supposed. And so, I use the image of a Mathland—where mathematics would become a natural vocabulary—to develop my idea that the computer presence could bring the humanistic and mathematical/scientific cultures together. In this book, Mathland is the first step in a larger argument about how the computer presence can change not only the way we teach children mathematics, but, much more fundamentally, the way in which our culture as a whole thinks about knowledge and learning.

To my ear the word "mathophobia" has two associations. One of these is a widespread fear of mathematics, which often has the intensity of a real phobia. The other comes from the meaning of the stem "math." In Greek it means "learning" in a general sense.* In

---

*The original meaning is present in the word "polymath," a person of many learnings. A less well-known word with the same stem which I shall use in later chapters is "mathetic," having to do with learning.

our culture, fear of learning is no less endemic (although more frequently disguised) than fear of mathematics. Children begin their lives as eager and competent learners. They have to *learn* to have trouble with learning in general and mathematics in particular. In both senses of "math" there is a shift from mathophile to mathophobe, from lover of mathematics and of learning to a person fearful of both. We shall look at how this shift occurs and develop some idea of how the computer presence could serve to counteract it. Let me begin with some reflections on what it is like to learn as a child.

That children learn a great deal seems so obvious to most people that they believe it is scarcely worth documenting. One area in which a high rate of learning is very plain is the acquisition of a spoken vocabulary. At age two very few children have more than a few hundred words. By the time they enter first grade, four years later, they know thousands of words. They are evidently learning many new words every day.

While we can "see" that children learn words, it is not quite as easy to see that they are learning mathematics at a similar or greater rate. But this is precisely what has been shown by Piaget's life-long study of the genesis of knowledge in children. One of the more subtle consequences of his discoveries is the revelation that adults fail to appreciate the extent and the nature of what children are learning, because knowledge structures we take for granted have rendered much of that learning invisible. We see this most clearly in what have come to be known as Piagetian "conservations" (see Figure 2).

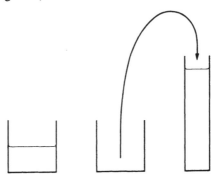

Figure 2   **The Conservation of Liquids**

For an adult it is obvious that pouring liquid from one glass to another does not change the volume (ignoring such little effects as drops that spilled or remained behind). The conservation of volume is so obvious that it seems not to have occurred to anyone before Piaget that children of four might not find it obvious at all.* A substantial intellectual growth is needed before children develop the "conservationist" view of the world. The conservation of volume is only one of many conservations they all learn. Another is the conservation of numbers. Again, it does not occur to most adults that a child must learn that counting a collection of objects in a different order should yield the same result. For adults counting is simply a method of determining how many objects "there are." The result of the operation is an "objective fact" independent of the act of counting. But the separation of number from counting (of product from process) rests on epistemological presuppositions not only unknown to preconservationist children, but alien to their worldview. These conservations are only part of a vast structure of "hidden" mathematical knowledge that children learn by themselves. In the intuitive geometry of the child of four or five, a straight line is not necessarily the shortest distance between two points, and walking slowly between two points does not necessarily take more time than walking fast. Here, too, it is not merely the "item" of knowledge that is missing, but the epistemological presupposition underlying the idea of "shortest" as a property of the path rather than of the action of traversing it.

None of this should be understood as mere *lack* of knowledge on the part of the children. Piaget has demonstrated how young children hold theories of the world that, in their own terms, are perfectly coherent. These theories, spontaneously "learned" by all children, have well-developed components that are not less "mathematical," though expressing a different mathematics, than the one generally accepted in our (adult) culture. The hidden learning process has at least two phases: Already in the preschool years every child first constructs one or more preadult theorizations of the

---

*People have lived with children for a long time. The fact that we had to wait for Piaget to tell us how children think and *what we all forget about our thinking as children* is so remarkable that it suggests a Freudian model of "cognitive repression."

world and then moves toward more adultlike views. And all this is done through what I have called Piagetian learning, a learning process that has many features the schools should envy: It is effective (all the children get there), it is inexpensive (it seems to require neither teacher nor curriculum development), and it is humane (the children seem to do it in a carefree spirit without explicit external rewards and punishments).

The extent to which adults in our society have lost the child's positive stance toward learning varies from individual to individual. An unknown but certainly significant proportion of the population has almost completely given up on learning. These people seldom, if ever, engage in deliberate learning and see themselves as neither competent at it nor likely to enjoy it. The social and personal cost is enormous: Mathophobia can, culturally and materially, limit people's lives. Many more people have not completely given up on learning but are still severely hampered by entrenched negative beliefs about their capacities. Deficiency becomes identity: "I can't learn French, I don't have an ear for languages;" "I could never be a businessman, I don't have a head for figures;" "I can't get the hang of parallel skiing, I never was coordinated." These beliefs are often repeated ritualistically, like superstitions. And, like superstitions, they create a world of taboos; in this case, taboos on learning. In this chapter and chapter 3, we discuss experiments that demonstrate that these self-images often correspond to a very limited reality—usually to a person's "school reality." In a learning environment with the proper emotional and intellectual support, the "uncoordinated" can learn circus arts like juggling and those with "no head for figures" learn not only that they can do mathematics but that they can enjoy it as well.

Although these negative self-images can be overcome, in the life of an individual they are extremely robust and powerfully self-reinforcing. If people believe firmly enough that they cannot do math, they will usually succeed in preventing themselves from doing whatever they recognize as math. The consequences of such self-sabotage is personal failure, and each failure reinforces the original belief. And such beliefs may be most insidious when held not only by individuals, but by our entire culture.

Our children grow up in a culture permeated with the idea that there are "smart people" and "dumb people." The social construction of the individual is as a bundle of aptitudes. There are people who are "good at math" and people who "can't do math." Everything is set up for children to attribute their first unsuccessful or unpleasant learning experiences to their own disabilities. As a result, children perceive failure as relegating them either to the group of "dumb people" or, more often, to a group of people "dumb at x" (where, as we have pointed out, x often equals mathematics). Within this framework children will define themselves in terms of their limitations, and this definition will be consolidated and reinforced throughout their lives. Only rarely does some exceptional event lead people to reorganize their intellectual self-image in such a way as to open up new perspectives on what is learnable.

This belief about the structure of human abilities is not easy to undermine. It is never easy to uproot popular beliefs. But here the difficulty is compounded by several other factors. First, popular theories about human aptitudes seem to be supported by "scientific" ones. After all, psychologists talk in terms of measuring aptitudes. But the significance of what is measured is seriously questioned by our simple thought experiment of imagining Mathland.

Although the thought experiment of imagining a Mathland leaves open the question of how a Mathland can actually be created, it is completely rigorous as a demonstration that the accepted beliefs about mathematical aptitude do not follow from the available evidence.[1] But since truly mathophobic readers might have trouble making this experiment their own, I shall reinforce the argument by casting it in another form. Imagine that children were forced to spend an hour a day drawing dance steps on squared paper and had to pass tests in these "dance facts" before they were allowed to dance physically. Would we not expect the world to be full of "dancophobes"? Would we say that those who made it to the dance floor and music had the greatest "aptitude for dance"? In my view, it is no more appropriate to draw conclusions about mathematical aptitude from children's unwillingness to spend many hundreds of hours doing sums.

One might hope that if we pass from parables to the more rigorous methods of psychology we could get some "harder" data on the problem of the true ceilings of competence attainable by individuals. But this is not so: The paradigm in use by contemporary educational psychology is focused on investigations of how children learn or (more usually) don't learn mathematics in the "anti-Mathland" in which we all live. The direction of such research has an analogy in the following parable:

> Imagine someone living in the nineteenth century who felt the need to improve methods of transportation. He was persuaded that the route to new methods started with a deep understanding of the existing problems. So he began a careful study of the differences among horse-drawn carriages. He carefully documented by the most refined methods how speed varied with the form and substance of various kinds of axles, bearings, and harnessing techniques.

In retrospect, we know that the road that led from nineteenth-century transportation was quite different. The invention of the automobile and the airplane did not come from a detailed study of how their predecessors, such as horse-drawn carriages, worked or did not work. Yet, this is the model for contemporary educational research. The standard paradigms for education research take the existing classroom or extracurricular culture as the primary object of study. There are many studies concerning the poor notions of math or science students acquire from today's schooling. There is even a very prevalent "humanistic" argument that "good" pedagogy should take these poor ways of thinking as its starting point. It is easy to sympathize with the humane intent. Nevertheless I think that the strategy implies a commitment to preserving the traditional system. It is analogous to improving the axle of the horse-drawn cart. But the real question, one might say, is whether we can invent the "educational automobile." Since this question (the central theme of this book) has not been addressed by educational psychology, we must conclude that the "scientific" basis for beliefs about aptitudes is really very shaky. But these beliefs are institutionalized in schools, in testing systems, and in college admissions criteria and consequently, their social basis is as firm as their scientific basis is weak.

From kindergarten on, children are tested for verbal and quantitative aptitudes, conceived of as "real" and separable entities. The results of these tests enter into the social construction of each child as a bundle of aptitudes. Once Johnny and his teacher have a shared perception of Johnny as a person who is "good at" art and "poor at" math, this perception has a strong tendency to dig itself in. This much is widely accepted in contemporary educational psychology. But there are deeper aspects to how school constructs aptitudes. Consider the case of a child I observed through his eighth and ninth years. Jim was a highly verbal and mathophobic child from a professional family. His love for words and for talking showed itself very early, long before he went to school. The mathophobia developed at school. My theory is that it came as a direct result of his verbal precocity. I learned from his parents that Jim had developed an early habit of describing in words, often aloud, whatever he was doing as he did it. This habit caused him minor difficulties with parents and preschool teachers. The real trouble came when he hit the arithmetic class. By this time he had learned to keep "talking aloud" under control, but I believe that he still maintained his inner running commentary on his activities. In his math class he was stymied: He simply did not know how to talk about doing sums. He lacked a vocabulary (as most of us do) and a sense of purpose. Out of this frustration of his verbal habits grew a hatred of math, and out of the hatred grew what the tests later confirmed as poor aptitude.

For me the story is poignant. I am convinced that what shows up as intellectual weakness very often grows, as Jim's did, out of intellectual strengths. And it is not only verbal strengths that undermine others. Every careful observer of children must have seen similar processes working in different directions: For example, a child who has become enamored of logical order is set up to be turned off by English spelling and to go on from there to develop a global dislike for writing.

The Mathland concept shows how to use computers as vehicles to escape from the situation of Jim and his dyslexic counterpart. Both children are victims of our culture's hard-edged separation between the verbal and the mathematical. In the Mathland we shall describe in this chapter, Jim's love and skill for language

45

could be mobilized to serve his formal mathematical development instead of opposing it, and the other child's love for logic could be recruited to serve the development of interest in linguistics.

The concept of mobilizing a child's multiple strengths to serve all domains of intellectual activity is an answer to the suggestion that differing aptitudes may reflect actual differences in brain development. It has become commonplace to talk as if there are separate brains, or separate "organs" in the brain, for mathematics and for language. According to this way of thinking, children split into the verbally and the mathematically apt depending on which brain organs are strongest. But the argument from anatomy to intellect reflects a set of epistemological assumptions. It assumes, for example, that there is only one route to mathematics and that if this route is "anatomically blocked," the child cannot get to the destination. Now, in fact, for most children in contemporary societies there may indeed be only one route into "advanced" mathematics, the route via school math. But even if further research in brain biology confirms that this route depends on anatomical brain organs that might be missing in some children, it would not follow that mathematics itself is dependent on these brain organs. Rather, it would follow that we should seek out other routes. Since this book is an argument that alternate routes do exist, it can be read as showing how the dependency of function on brain is itself a social construct.

Let us grant, for the sake of argument, that there is a special part of the brain especially good at performing the mental manipulations of numbers we teach children in school, and let's call it the MAD, or "math acquisition device."[2] On this assumption it would make sense that in the course of history humankind would have evolved methods of doing and of teaching arithmetic that take full advantage of the MAD. But while these methods would work for most of us, and so for society as a whole, reliance on them would be catastrophic for an individual whose MAD happened to be damaged or inaccessible for some other (perhaps "neurotic") reason. Such a person would fail at school and be diagnosed as a victim of "dyscalculia." And as long as we insist on making children learn arithmetic by the standard route, we will continue to "prove" by objective tests that these children really cannot "do arithmetic." But this is like proving that the deaf children cannot have language because they don't hear. Just as sign languages use hands and eyes to bypass the more usual speaking organs so, too, alternative ways of doing mathematics that bypass the MAD may be as good as, even if different from, the usual ones.

But we do not have to appeal to neurology to explain why some children do not become fluent in mathematics. The analogy of the dance class without music or dance floor is a serious one. Our education culture gives mathematics learners scarce resources for making sense of what they are learning. As a result our children are forced to follow the very worst model for learning mathematics. This is the model of rote learning, where material is treated as meaningless; it is a *dissociated* model. Some of our difficulties in teaching a more culturally integrable mathematics have been due to an objective problem: Before we had computers there were very few good points of contact between what is most fundamental and engaging in mathematics and anything firmly planted in everyday life. But the computer—a mathematics-speaking being in the midst of the everyday life of the home, school, and workplace—is able to provide such links. The challenge to education is to find ways to exploit them.

Mathematics is certainly not the only example of dissociated learning. But it is a very good example for precisely the reason that many readers are probably now wishing that I would talk about something else. Our culture is so mathophobic, so math-fearing, that if I could demonstrate how the computer can bring us into a new relationship to mathematics, I would have a strong foundation for claiming that the computer has the ability to change our relation to other kinds of learning we might fear. Experiences in Mathland, such as entering into a "mathematical conversation," give the individual a liberating sense of the possibilities of doing a variety of things that may have previously seemed "too hard." In this sense, contact with the computer can open access to knowledge for people, not instrumentally by providing them with processed information, but by challenging some constraining assumptions they make about themselves.

The computer-based Mathland I propose extends the kind of natural, Piagetian learning that accounts for children's learning a first language to learning mathematics. Piagetian learning is typically deeply embedded in other activities. For example, the infant does not have periods set aside for "learning talking." This model of learning stands in opposition to dissociated learning, learning that takes place in relative separation from other kinds of activities, mental and physical. In our culture, the teaching of mathematics in schools is paradigmatic of dissociated learning. For most people, mathematics is taught and taken as medicine. In its dissociation of mathematics, our culture comes closest to caricaturing its own worst habits of epistemological alienation. In LOGO environments we have done some blurring of boundaries: No particular computer activities are set aside as "learning mathematics."

The problem of making mathematics "make sense" to the learner touches on the more general problem of making a language of "formal description" make sense. So before turning to examples of how the computer helps give meaning to mathematics, we shall look at several examples where the computer helped give meaning to a language of formal description in domains of knowledge that people do not usually count as mathematics. In our first example the domain is grammar, for many people a subject only a little less threatening than math.

Well into a year-long study that put powerful computers in the classrooms of a group of "average" seventh graders, the students were at work on what they called "computer poetry." They were using computer programs to generate sentences. They gave the computer a syntactic structure within which to make random choices from given lists of words. The result is the kind of concrete poetry we see in the illustration that follows. One of the students, a thirteen-year-old named Jenny, had deeply touched the project's staff by asking on the first day of her computer work, "Why were we chosen for this? We're not the brains." The study had deliberately chosen children of "average" school performance. One day Jenny came in very excited. She had made a discovery. "Now I know why we have nouns and verbs," she said. For many years in school Jenny had been drilled in grammatical categories. She had never understood the differences between nouns and verbs and adverbs. But now it was apparent that her difficulty with grammar was not due to an inability to work with logical categories. It was something else. She had simply seen no purpose in the enterprise. She had not been able to make any sense of what grammar was about in the sense of what it might be *for*. And when she had asked what it was for, the explanations that her teachers gave seemed manifestly dishonest. She said she had been told that "grammar helps you talk better."

INSANE RETARD MAKES BECAUSE SWEET SNOOPY SCREAMS
SEXY WOLF LOVES THATS WHY THE SEXY LADY HATES
UGLY MAN LOVES BECAUSE UGLY DOG HATES
MAD WOLF HATES BECAUSE INSANE WOLF SKIPS
SEXY RETARD SCREAMS THATS WHY THE SEXY RETARD HATES
THIN SNOOPY RUNS BECAUSE FAT WOLF HOPS
SWEET FOGINY SKIPS A FAT LADY RUNS

Jenny's Concrete Poetry

In fact, tracing the connection between learning grammar and improving speech requires a more distanced view of the complex process of learning language than Jenny could have been given at the age she first encountered grammar. She certainly didn't see any way in which grammar could help talking, nor did she think her talking needed any help. Therefore she learned to approach grammar with resentment. And, as is the case for most of us, resentment guaranteed failure. But now, as she tried to get the computer to generate poetry, something remarkable happened. She found herself classifying words into categories, not because she had been told she had to but because she needed to. In order to "teach" her computer to make strings of words that would look like English, she had to "teach" it to choose words of an appropriate class. What she learned about grammar from this experience with a machine was anything but mechanical or routine. Her learning was deep and meaningful. Jenny did more than learn definitions for particular

49

grammatical classes. She understood the general idea that words (like things) can be placed in different groups or sets, and that doing so could work for her. She not only "understood" grammar, she changed her relationship to it. It was "hers," and during her year with the computer, incidents like this helped Jenny change her image of herself. Her performance changed too; her previously low to average grades became "straight A's" for her remaining years of school. She learned that she could be "a brain" after all.

It is easy to understand why math and grammar fail to make sense to children when they fail to make sense to everyone around them and why helping children to make sense of them requires more than a teacher making the right speech or putting the right diagram on the board. I have asked many teachers and parents what they thought mathematics to be and why it was important to learn it. Few held a view of mathematics that was sufficiently coherent to justify devoting several thousand hours of a child's life to learning it, and children sense this. When a teacher tells a student that the reason for those many hours of arithmetic is to be able to check the change at the supermarket, the teacher is simply not believed. Children see such "reasons" as one more example of adult double talk. The same effect is produced when children are told school math is "fun" when they are pretty sure that teachers who say so spend their leisure hours on anything except this allegedly fun-filled activity. Nor does it help to tell them that they need math to become scientists—most children don't have such a plan. The children can see perfectly well that the teacher does not like math any more than they do and that the reason for doing it is simply that it has been inscribed into the curriculum. All of this erodes children's confidence in the adult world and the process of education. *And I think it introduces a deep element of dishonesty into the educational relationship.*

Children perceive the school's rhetoric about mathematics as double talk. In order to remedy the situation we must first acknowledge that the child's perception is fundamentally correct. The *kind of mathematics* foisted on children in schools is not meaningful, fun, or even very useful. This does not mean that an individual child cannot turn it into a valuable and enjoyable personal game.

For some the game is scoring grades; for others it is outwitting the teacher and the system. For many, school math is enjoyable in its repetitiveness, precisely because it is so mindless and dissociated that it provides a shelter from having to think about what is going on in the classroom. But all this proves is the ingenuity of children. It is not a justification for school math to say that *despite* its intrinsic dullness, inventive children can find excitement and meaning in it.

It is important to remember the distinction between *mathematics*—a vast domain of inquiry whose beauty is rarely suspected by most nonmathematicians—and something else which I shall call *math* or *school math*.

I see "school math" as a social construction, a kind of QWERTY. A set of historical accidents (which shall be discussed in a moment) determined the choice of certain mathematical topics as *the* mathematical baggage that citizens should carry. Like the QWERTY arrangement of typewriter keys, school math did make some sense in a certain historical context. But, like QWERTY, it has dug itself in so well that people take it for granted and invent rationalizations for it long after the demise of the historical conditions that made sense of it. Indeed, for most people in our culture it is inconceivable that school math could be very much different: This is the only mathematics they know. In order to break this vicious circle I shall lead the reader into a new area of mathematics, Turtle geometry, that my colleagues and I have created as a better, more meaningful first area of formal mathematics for children. The design criteria of Turtle geometry are best understood by looking a little more closely at the historical conditions responsible for the shape of school math.

Some of these historical conditions were pragmatic. Before electronic calculators existed it was a practical social necessity that many people be "programmed" to perform such operations as long division quickly and accurately. But now that we can purchase calculators cheaply we should reconsider the need to expend several hundred hours of every child's life on learning such arithmetic functions. I do not mean to deny the intellectual value of some knowledge, indeed, of a lot of knowledge, about numbers. Far from

51

it. But we can now select this knowledge on coherent, rational grounds. We can free ourselves from the tyranny of the superficial, pragmatic considerations that dictated past choices about what knowledge should be learned and at what age.

But utility was only one of the historical reasons for school math. Others were of a *mathetic* nature. Mathetics is the set of guiding principles that govern learning. Some of the historical reasons for school math had to do with what was learnable and teachable in the precomputer epoch. As I see it, a major factor that determined what mathematics went into school math was what could be done in the setting of school classrooms with the primitive technology of pencil and paper. For example, children can draw graphs with pencil and paper. So it was decided to let children draw many graphs. The same considerations influenced the emphasis on certain kinds of geometry. For example, in school math "analytic geometry" has become synonymous with the representation of curves by equations. As a result every educated person vaguely remembers that $y = x^2$ is the equation of a parabola. And although most parents have very little idea of why anyone should know this, they become indignant when their children do not. They assume that there must be a profound and objective reason known to those who better understand these things. Ironically, their mathophobia keeps most people from trying to examine those reasons more deeply and thus places them at the mercy of the self-appointed math specialists. Very few people ever suspect that the reason for what is included and what is not included in school math might be as crudely technological as the ease of production of parabolas with pencils! This is what could change most profoundly in a computer-rich world: The range of easily produced mathematical constructs will be vastly expanded.

Another mathetic factor in the social construction of school math is the technology of grading. A living language is learned by speaking and does not need a teacher to verify and grade each sentence. A dead language requires constant "feedback" from a teacher. The activity known as "sums" performs this feedback function in school math. These absurd little repetitive exercises have only one merit: They are easy to grade. But this merit has bought them a firm place at the center of school math. In brief, I maintain that construction of school math is strongly influenced by what seemed to be teachable when math was taught as a "dead" subject, using the primitive, passive technologies of sticks and sand, chalk and blackboard, pencil and paper. The result was an intellectually incoherent set of topics that violates the most elementary mathetic principles of what makes certain material easy to learn and some almost impossible.

Faced with the heritage of school, math education can take two approaches. The traditional approach accepts school math as a given entity and struggles to find ways to teach it. Some educators use computers for this purpose. Thus, paradoxically, the most common use of the computer in education has become force-feeding indigestible material left over from the precomputer epoch. In Turtle geometry the computer has a totally different use. There the computer is used as a mathematically expressive medium, one that frees us to design personally meaningful and intellectually coherent and easily learnable mathematical topics for children. Instead of posing the educational problem as "how to teach the existing school math," we pose it as "reconstructing mathematics," or more generally, as reconstructing knowledge in such a way that no great effort is needed to teach it.

All "curriculum development" could be described as "reconstructing knowledge." For example, the New Math curriculum reform of the sixties made some attempt to change the content of school math. But it could not go very far. It was stuck with having to do sums, albeit different sums. The fact that the new sums dealt with sets instead of numbers, or arithmetic in base two instead of base ten made little difference. Moreover, the math reform did not provide a challenge to the inventiveness of creative mathematicians and so never acquired the sparkle of excitement that marks the product of new thought. The name itself—"New Math"—was a misnomer. There was very little new about its mathematical content: It did not come from a process of invention of children's mathematics but from a process of trivialization of mathematician's mathematics. Children need and deserve something better than selecting out pieces of old mathematics. Like clothing passed down to the younger siblings, it never fits comfortably.

Turtle geometry started with the goal of fitting children. Its primary design criterion was to be *appropriable*. Of course it had to have serious mathematical content, but we shall see that appropriability and serious mathematic thinking are not at all incompatible. On the contrary: We shall end up understanding that some of the most personal knowledge is also the most profoundly mathematical. In many ways mathematics—for example the mathematics of space and movement and repetitive patterns of action—is what comes most naturally to children. It is into this mathematics that we sink the tap-root of Turtle geometry. As my colleagues and I have worked through these ideas, a number of principles have given more structure to the concept of an appropriable mathematics. First, there was the *continuity principle*: The mathematics must be continuous with well-established personal knowledge from which it can inherit a sense of warmth and value as well as "cognitive" competence. Then there was the *power principle*: It must empower the learner to perform personally meaningful projects that could not be done without it. Finally there was a *principle of cultural resonance*: The topic must make sense in terms of a larger social context. I have spoken of Turtle geometry making sense to children. But it will not truly make sense to children unless it is accepted by adults too. A dignified mathematics for children cannot be something we permit ourselves to inflict on children, like unpleasant medicine, although we see no reason to take it ourselves.

Chapter 3

Turtle Geometry: A Mathematics Made for Learning

TURTLE GEOMETRY is a different style of doing geometry, just as Euclid's axiomatic style and Descartes's analytic style are different from one another. Euclid's is a *logical* style. Descartes's is an *algebraic* style. Turtle geometry is a *computational* style of geometry.

Euclid built his geometry from a set of fundamental concepts, one of which is the point. A point can be defined as an entity that has a position but no other properties—it has no color, no size, no shape. People who have not yet been initiated into formal mathematics, who have not yet been "mathematized," often find this notion difficult to grasp, and even bizarre. It is hard for them to relate it to anything else they know. Turtle geometry, too, has a fundamental entity similar to Euclid's point. But this entity, which I call a "Turtle," can be related to things people know because unlike Euclid's point, it is not stripped so totally of all properties, and instead of being static it is dynamic. Besides position the Turtle has one other important property: It has "heading." A Euclidean point is at some place—it has a position, and that is all you can say about it. A Turtle is at some place—it, too, has a position—but it also faces some direction—its heading. In this, the Turtle is like a person—I am *here* and I am facing north—or an animal or a boat.

And from these similarities comes the Turtle's special ability to serve as a first representative of formal mathematics for a child. Children can *identify* with the Turtle and are thus able to bring their knowledge about their bodies and how they move into the work of learning formal geometry.

To see how this happens we need to know one more thing about Turtles: They are able to accept commands expressed in a language called TURTLE TALK. The command FORWARD causes the Turtle to move in a straight line in the direction it is facing (see Figure 3). To tell it how far to go, FORWARD must be followed by a number: FORWARD 1 will cause a very small movement, FORWARD 100 a larger one. In LOGO environments many children have been started on the road to Turtle geometry by introducing them to a mechanical turtle, a cybernetic robot, that will carry out these commands when they are typed on a typewriter keyboard. This "floor Turtle" has wheels, a dome shape, and a pen so that it can draw a line as it moves. But its essential properties—position, heading, and ability to obey TURTLE TALK commands—are the ones that matter for doing geometry. The child may later meet these same three properties in another embodiment of the Turtle: a "Light Turtle." This is a triangular-shaped object on a television screen. It too has a position and a heading. And it too moves in response to the same TURTLE TALK commands. Each kind of Turtle has its strong points: The floor Turtle can be used as a bulldozer as well as a drawing instrument; the Light Turtle draws bright-colored lines faster than the eye can follow. Neither is better, but the fact that there are two carries a powerful idea: Two *physically* different entities can be *mathematically* the same (or "isomorphic").[1]

The commands FORWARD and BACK cause a Turtle to move in a straight line in the direction of its heading: Its position changes, but its heading remains the same. Two other commands change the heading without affecting the position: RIGHT and LEFT cause a Turtle to "pivot," to change heading while remaining in the same place. Like FORWARD, a turning command also needs to be given a number—an input message—to say how much the Turtle should turn. An adult will quickly recognize these numbers as the measure of the turning angle in degrees. For most children these numbers have to be explored, and doing so is an exciting and playful process.

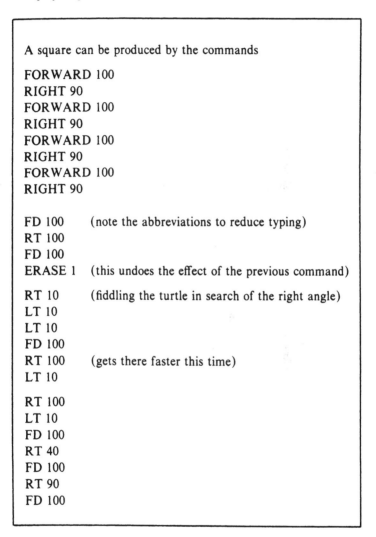

A square can be produced by the commands

FORWARD 100
RIGHT 90
FORWARD 100
RIGHT 90
FORWARD 100
RIGHT 90
FORWARD 100
RIGHT 90

FD 100   (note the abbreviations to reduce typing)
RT 100
FD 100
ERASE 1   (this undoes the effect of the previous command)

RT 10   (fiddling the turtle in search of the right angle)
LT 10
LT 10
FD 100
RT 100   (gets there faster this time)
LT 10

RT 100
LT 10
FD 100
RT 40
FD 100
RT 90
FD 100

Figure 3
An Actual Transcript of a Child's Early Attempt at a Square

Since learning to control the Turtle is like learning to speak a language it mobilizes the child's expertise and pleasure in speaking. Since it is like being in command, it mobilizes the child's expertise and pleasure in commanding. To make the Turtle trace a square you walk in a square yourself and describe what you are doing in TURTLE TALK. And so, working with the Turtle mobilizes the child's expertise and pleasure in motion. It draws on the child's well-established knowledge of "body-geometry" as a starting point for the development of bridges into formal geometry.

The goal of children's first experiences in the Turtle learning environment is not to learn formal rules but to develop insights into the way they move about in space. These insights are described in TURTLE TALK and thereby become "programs" or "procedures" or "differential equations" for the Turtle. Let's look closely at how a child, who has already learned to move the Turtle in straight lines to draw squares, triangles, and rectangles, might learn how to program it to draw a circle.

Let us imagine, then, as I have seen a hundred times, a child who demands: How can I make the Turtle draw a circle? The instructor in a LOGO environment does not provide answers to such questions but rather introduces the child to a method for solving not only this problem but a large class of others as well. This method is summed up in the phrase "play Turtle." The child is encouraged to move his or her body as the Turtle on the screen must move in order to make the desired pattern. For the child who wanted to make a circle, moving in a circle might lead to a description such as: "When you walk in a circle you take a little step forward and you turn a little. And you keep doing it." From this description it is only a small step to a formal Turtle program.

TO CIRCLE REPEAT [FORWARD 1 RIGHT 1]

Another child, perhaps less experienced in simple programming and in the heuristics of "playing Turtle," might need help. But the help would not consist primarily of teaching the child how to program the Turtle circle, but rather of teaching the child a method, a heuristic procedure. This method (which includes the advice summed up as "play Turtle") tries to establish a firm connection between personal activity and the creation of formal knowledge.

In the Turtle Mathland anthropomorphic images facilitate the transfer of knowledge from familiar settings to new contexts. For example, the metaphor for what is usually called "programming computers" is teaching the Turtle a new word. A child who wishes to draw many squares can teach the Turtle a new command that will make it carry out in sequence the seven commands used to draw a square as is shown in Figure 3. This can be given to the computer in several different forms among which are:

```
TO SQUARE
FORWARD 100
RIGHT 90
FORWARD 100
RIGHT 90
FORWARD 100
RIGHT 90
FORWARD 100
END

TO SQUARE
REPEAT 4
  FORWARD 100
  RIGHT 90
END

TO SQUARE :SIZE
REPEAT 4
  FORWARD :SIZE
  RIGHT 90
END
```

Similarly we can program an equilateral triangle by:

```
TO TRIANGLE
FORWARD 100
RIGHT 120
FORWARD 100
RIGHT 120
FORWARD 100
END

TO TRIANGLE :SIDE
REPEAT 3
  FORWARD :SIDE
  RIGHT 120
END
```

These alternative programs achieve almost the same effects but informed readers will notice some differences. The most obvious difference is in the fact that some of them allow figures to be drawn with different sizes: In these cases the command to draw the figure would have to be SQUARE 50 or SQUARE 100 rather than simply SQUARE. A more subtle difference is in the fact that some of them leave the Turtle in its original state. Programs written in this clean style are much easier to understand and use in a variety of contexts. And in noticing this difference children learn two kinds of lessons. They learn a general "mathetic principle," making components to favor modularity. And they learn to use the very powerful idea of "state."

The same strategy of moving from the familiar to the unknown brings the learner into touch with some powerful general ideas: for example, the idea of hierarchical organization (of knowledge, of organizations, and of organisms), the idea of planning in carrying through a project, and the idea of debugging.

One does not need a computer to draw a triangle or a square. Pencil and paper would do. But once these programs have been constructed they become building blocks that enable a child to create hierarchies of knowledge. Powerful intellectual skills are developed in the process—a point that is most clearly made by looking at some projects children have set for themselves after a few sessions with the Turtle. Many children have spontaneously followed the same path as Pamela. She began by teaching the compute-

SQUARE and TRIANGLE as described previously. Now she saw that she can build a house by putting the triangle on top of the square. So she tries:

TO HOUSE
SQUARE
TRIANGLE
END

But when she gives the command HOUSE, the Turtle draws (Figure 4.):

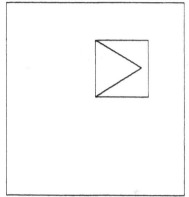

Figure 4

The triangle came out inside the square instead of on top of it!

Typically in math class, a child's reaction to a wrong answer is to try to *forget* it as fast as possible. But in the LOGO environment, the child is not criticized for an error in drawing. The process of debugging is a normal part of the process of understanding a program. The programmer is encouraged to study the bug rather than forget the error. And in the Turtle context there is a good reason to study the bug. It will pay off.

There are many ways this bug can be fixed. Pamela found one of them by playing Turtle. By walking along the Turtle's track she saw that the triangle got inside the square because its first turning move in starting the triangle was a right turn. So she could fix the bug by making a left-turning triangle program. Another common way to fix this bug is by inserting a RIGHT 30 between SQUARE and TRIANGLE. In either case the amended procedure makes the following picture (Figure 5).

61

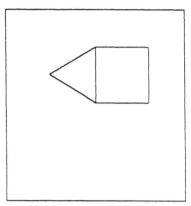

Figure 5

The learner sees progress, and also sees that things are not often either completely right or completely wrong but, rather, are on a continuum. The house is better but still has a bug. With a little more playing Turtle this final bug is pinned down and fixed by doing a RIGHT 90 as the first step in the program.

Some children use program building blocks to make concrete drawings such as HOUSE. Others prefer more abstract effects. For example, if you give the command SQUARE, pivot the Turtle with a RIGHT 120, do SQUARE again, pivot the TURTLE with RT 120 or with RT 10, do SQUARE once more and keep repeating, you get the picture in Figure 6a. A smaller rotation gives the picture in Figure 6b.

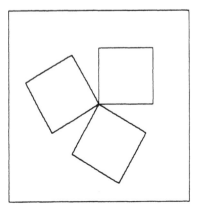

Figure 6a

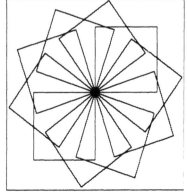

Figure 6b

Turtle Geometry: A Mathematics Made for Learning

These examples show how the continuity and the power principles make Turtle geometry learnable. But we wanted it to do something else as well, to open intellectual doors, preferably to be a carrier of important, powerful ideas. Even in drawing these simple squares and stars the Turtle carried some important ideas: angle, controlled repetition, state-change operator. To give ourselves a more systematic overview of what children learn from working with the Turtle we begin by distinguishing between two kinds of knowledge. One kind is *mathematical*: The Turtles are only a small corner of a large mathematical subject, Turtle geometry, a kind of geometry that is easily learnable and an effective carrier of very general mathematical ideas. The other kind of knowledge is *mathetic*: knowledge about learning. First we shall look more closely at the mathetic aspects of the Turtle experience and then turn to its more technically mathematical side. Of course, the two overlap.

We introduced Turtle geometry by relating it to a fundamental mathetic principle: Make sense of what you want to learn. Recall the case of Jenny, who possessed the conceptual prerequisites for defining nouns or verbs but who could not learn grammar because she could not *identify with* this enterprise. In this very fundamental way grammar did not make sense to her. Turtle geometry was specifically designed to be something children *could* make sense of, to be something that would resonate with their sense of what is important. And it was designed to help children develop the mathetic strategy: In order to learn something, first make sense of it.

The Turtle circle incident illustrates *syntonic learning*.[2] This term is borrowed from clinical psychology and can be contrasted to the dissociated learning already discussed. Sometimes the term is used with qualifiers that refer to kinds of syntonicity. For example, the Turtle circle is *body syntonic* in that the circle is firmly related to children's sense and knowledge about their own bodies. Or it is *ego syntonic* in that it is coherent with children's sense of themselves as people with intentions, goals, desires, likes, and dislikes. A child who draws a Turtle circle wants to draw the circle; doing it produces pride and excitement.

Turtle geometry is learnable because it is syntonic. And it is an aid to learning other things because it encourages the conscious, deliberate use of problem-solving and mathetic strategies. Mathematician George Polya[3] has argued that general methods for solving problems should be taught. Some of the strategies used in Turtle geometry are special cases of Polya's suggestions. For example, Polya recommends that whenever we approach a problem we should run through a mental checklist of heuristic questions such as: Can this problem be subdivided into simpler problems? Can this problem be related to a problem I already know how to solve? Turtle geometry lends itself to this exercise. The key to finding out how to make a Turtle draw a circle is to refer to a problem whose solution is known very well indeed—the problem of walking in a circle. Turtle geometry provides excellent opportunities to practice the art of splitting difficulties. For example, HOUSE was made by first making SQUARE and TRIANGLE. In short, I believe that Turtle geometry lends itself so well to Polya's principles that the best way to explain Polya to students is to let them learn Turtle geometry. Thus, Turtle geometry serves as a carrier for the general ideas of a heuristic strategy.

Because of Polya's influence, it has often been suggested that mathematics teachers pay explicit attention to heuristics or "process" as well as to content. The failure of this idea to take root in the educational system can be explained partially by the paucity of good situations in which simple and compelling models of heuristic knowledge can be encountered and internalized by children. Turtle geometry is not only particularly rich in such situations, it also adds a new element to Polya's advice: To solve a problem look for something like it that you already understand. The advice is abstract; Turtle geometry turns it into a concrete, procedural principle: *Play Turtle. Do it yourself.* In Turtle work an almost inexhaustible source of "similar situations" is available because we draw on our own behavior, our own bodies. So, when in trouble, we can play Turtle. This brings Polya's advice down to earth. Turtle geometry becomes a bridge to Polya. The child who has worked extensively with Turtles becomes deeply convinced of the value of "looking for something like it" because the advice has often paid off. From these successes comes the confidence and skill needed to learn how to apply the principle in situations, such as most of those encountered in school math, where similarities are less evident. School math, though elementary in terms of its arithmetic content, is a relatively advanced subject for the exercise of Polya's principles.

Arithmetic is a bad introductory domain for learning heuristic thinking. Turtle geometry is an excellent one. By its qualities of ego and body syntonicity, the act of learning to make the Turtle draw gives the child a model of learning that is very much different from the dissociated one a fifth-grade boy, Bill, described as the way to learn multiplication tables in school: "You learn stuff like that by making your mind a blank and saying it over and over until you know it." Bill spent a considerable amount of time on "learning" his tables. The results were poor and, in fact, the poor results themselves speak for the accuracy of Bill's reporting of his own mental processes in learning. He failed to learn because he forced himself out of any relationship to the material—or rather, he adopted the worst relationship, dissociation, as a strategy for learning. His teachers thought that he "had a poor memory" and had even discussed the possibility of brain damage. But Bill had extensive knowledge of popular and folk songs, which he had no difficulty remembering, perhaps because he was too busy to think about making his mind a blank.

Current theories about the separation of brain functions might suggest that Bill's "poor memory" was specific to numbers. But the boy could easily recount reference numbers, prices, and dates for thousands of records. The difference between what he "could" and "could not" learn did not depend on the content of the knowledge but on his relationship to it. Turtle geometry, by virtue of its connection with rhythm and movement and the navigational knowledge needed in everyday life, allowed Billy to relate to it more as he did to songs than to multiplication tables. His progress was spectacular. Through Turtle geometry, mathematical knowledge Billy had previously rejected could enter his intellectual world.

Now we turn from mathetic to mathematical considerations. What mathematics does one learn when one learns Turtle geometry? For the purposes of this discussion we distinguish three classes of mathematical knowledge, each of which benefits from work with

Turtles. First, there is the body of knowledge "school math" that has been explicitly selected (in my opinion largely by historical accident) as *the* core of basic mathematics that all citizens should possess. Second, there is a body of knowledge (let me call it "protomath") that is presupposed by school math even though it is not explicitly mentioned in traditional curricula. Some of this knowledge is of a general "social" nature: for example, knowledge that bears on why we do mathematics at all and how we can make sense of math. Other knowledge in this category is the kind of underlying structure to which genetic epistemology has drawn the attention of educators: deductive principles such as transitivity, the conservations, the intuitive logic of classifications, and so on. Finally, there is a third category: knowledge that is neither included in nor presupposed by the school math but that *ought* to be considered for inclusion in the intellectual equipment of the educated citizen of the future.

I think that understanding the relations among the Euclidean, the Cartesian, and the differential systems of geometry belongs to this third category. For a student, drawing a Turtle circle is more than a "common sense" way of drawing circles. It places the child in contact with a cluster of ideas that lie at the heart of the calculus. This fact may be invisible to many readers whose only encounter with calculus was a high school or college course where "calculus" was equated with certain formal manipulations of symbols. The child in the Turtle circle incident was not learning about the *formalism* of calculus, for example that the derivative of x^n is nx^{n-1}, but about its use and its *meaning*. In fact the Turtle circle program leads to an alternative formalism for what is traditionally called a "differential equation" and is a powerful carrier of the ideas behind the differential. This is why it is possible to understand so many topics through the Turtle; *the Turtle program is an intuitive analog of the differential equation, a concept one finds in almost every example of traditional applied mathematics.*

Differential calculus derives much of its power from an ability to describe growth by what is happening at the growing tip. This is what made it such a good instrument for Newton's attempts to understand the motion of the planets. As the orbit is traced out, it is the local conditions at the place where the planet now finds itself that determine where it will go next. In our instructions to the Turtle, FORWARD 1, RIGHT TURN 1, we referred only to the difference between where the Turtle is now and where it shall momentarily be. This is what makes the instructions *differential*. There is no reference in this to any distant part of space outside of the path itself. The Turtle sees the circle as it goes along, from within, as it were, and is blind to anything far away from it. This property is so important that mathematicians have a special name for it: Turtle geometry is "intrinsic." The spirit of intrinsic differential geometry is seen by looking at several ways to think about a curve, say, the circle. For Euclid, the defining characteristic of a circle is the constant distance between points on the circle and a point, the center, that is not itself part of the circle. In Descartes's geometry, in this respect more like Euclid's than that of the Turtle, points are situated by their distance from something outside of them, that is to say the perpendicular coordinate axes. Lines and curves are defined by equations connecting these coordinates. So, for example, a circle is described as:

$$(x-a)^2 + (y-b)^2 = R^2$$

In Turtle geometry a circle is defined by the fact that the Turtle keeps repeating the act: FORWARD a little, TURN a little. This repetition means that the curve it draws will have "constant curvature," where curvature means how much you turn for a given forward motion.[4]

Turtle geometry belongs to a family of geometries with properties not found in the Euclidean or Cartesian system. These are the differential geometries that have developed since Newton and have made possible much of modern physics. We have noted that the *differential equation* is the formalism through which physics has been able to describe the motion of a particle or a planet. In chapter 5, where we discuss this in more detail, we shall also see that it is the appropriate formalism to describe the motion of an animal or the evolution of an economy. And we shall come to understand more clearly that it is not by coincidence that Turtle geometry has links both to the experience of a child and to the most powerful achievements in physics. For the laws of motion of the child, though less precise in form, share the mathematical structure of the differential equation with the laws of motion of planets turning about the sun and with those of moths turning about a candle flame. And the Turtle is nothing more or less than a reconstruction in intuitive computational form of the qualitative core of this mathematical structure. When we return to these ideas in chapter 5, we shall see how Turtle geometry opens the door to an intuitive grasp of calculus, physics, and mathematical modeling as it is used in the biological and social sciences.

The effect of work with Turtle geometry on some components of school math is primarily *relational* or *affective:* Many children have come to the LOGO lab hating numbers as alien objects and have left loving them. In other cases work with the Turtle provides specific intuitive models for complex mathematical concepts most children find difficult. The use of numbers to measure angles is a simple example. In the Turtle context children pick this ability up almost unconsciously. Everyone—including the few first graders and many third graders we have worked with—emerges from the experience with a much better sense of what is meant by 45 degrees or 10 degrees or 360 degrees than the majority of high school students ever acquire. Thus, they are prepared for all the many formal topics—geometry, trigonometry, drafting, and so on—in which the concept of *angle* plays a central part. But they are prepared for something else as well, an aspect of the use of angular measure in our society to which the school math is systematically blind.

One of the most widespread representations of the idea of angle in the lives of contemporary Americans is in navigation. Many millions navigate boats or airplanes or read maps. For most there is a total dissociation between these *live* activities and the *dead* school math. We have stressed the fact that using the Turtle as metaphorical carrier for the idea of angle connects it firmly to body geometry. We have called this body syntonicity. Here we see a *cultural syntonicity:* The Turtle connects the idea of angle to navigation, activity firmly and positively rooted in the extraschool culture of many children. And as computers continue to spread into the world, the cultural syntonicity of Turtle geometry will become more and more powerful.

A second key mathematical concept whose understanding is facilitated by the Turtle is the idea of a *variable:* the idea of using a symbol to name an unknown entity. To see how Turtles contribute to this, we extend the program for Turtle circles into a program for Turtle spirals (Figure 7).

Figure 7

Look, for example, at the coil spiral. Like the circle, it too can be made according to the prescription: Go forward a little, turn a little. The difference between the two is that the circle is "the same all the way" while the spiral gets flatter, "less curvy," as you move out from the middle. The circle is a curve of constant curvature. The curvature of the spiral decreases as it moves outward. To walk in a spiral one could take a step, then turn, take a step, then turn, each time turning a little less (or stepping a little more). To translate this into instructions for the Turtle, you need some way to express the fact that you are dealing with a *variable* quantity. In principle you could describe this by a very long program (see Figure 8) that would specify precisely how much the Turtle should turn on each step. This is tedious. A better method uses the concept of symbolic naming through a variable, one of the most powerful mathematical ideas ever invented.

TO SPI	TO COIL
FORWARD 10	FORWARD 5
RIGHT 90	RIGHT 5
FORWARD 15	FORWARD 5
RIGHT 90	RIGHT 5 * .95
FORWARD 20	FORWARD 5
RIGHT 90	RIGHT 5 * .95 * .95
FORWARD 25	FORWARD 5
RIGHT 90	RIGHT 5 * .95 * .95 * .95
FORWARD 30	FORWARD 5
RIGHT 90	RIGHT 5 * .95 * .95 * .95
FORWARD 35	FORWARD 5
RIGHT 90	RIGHT 5 * .95 * .95 * .95 * .95
FORWARD 40	FORWARD 5
RIGHT 90	RIGHT 5 * .95 * .95 * .95 * .95 * .95
FORWARD 45	FORWARD 5
RIGHT 90	RIGHT 5 * .95 * .95 * .95 * .95 * .95 * .95
FORWARD 50	FORWARD 5
RIGHT 90	RIGHT 5 * .95 * .95 * .95 * .95 * .95 * .95 * .95
FORWARD 55	FORWARD 5
RIGHT 90	RIGHT 5 * .95 * .95 * .95 * .95 * .95 * .95 * .95 * .95
FORWARD 60	FORWARD 5
RIGHT 90	
FORWARD 65	etc.
RIGHT 90	

etc.

Figure 8
How NOT to Draw Spirals

In TURTLE TALK, variables are presented as a means of communication. What we want to say to the Turtle is "go forward a little step, then turn a certain amount, but I can't tell you now how much to turn because it will be different each time." To draw the "squiral" we want to say "go forward a certain distance, which will be different each time, and then turn 90." In mathematical language the trick for saying something like this is to invent a name for the "amount I can't tell you." The name could be a letter, such as X, or it could be a whole word, such as *ANGLE OR DISTANCE*. (One of the minor contributions of the computer culture to mathematics is its habit of using mnemonic words instead of single letters as names for variables.) To put the idea of variable to work, TURTLE TALK allows one to create a "procedure with an input." This can be done by typing:

```
TO STEP DISTANCE
FORWARD DISTANCE
RIGHT 90
END
```

The command STEP 100 will make the Turtle go forward 100 units and then turn right 90 degrees. Similarly STEP 110 will make it go forward 110 units and then turn 90 degrees. In LOGO environments we encourage children to use an anthropomorphic metaphor: The command STEP invokes an agent (a "STEP man") whose job is to issue two commands, a FORWARD command and a RIGHT command, to the Turtle. But this agent cannot perform this job without being given a message—a number that will be passed on to the "FORWARD man" who will pass it on to the Turtle.

The procedure STEP is not really very exciting, but a small change will make it so. Compare it with the procedure SPI, which is exactly the same except for having one extra line:

```
TO SPI DISTANCE
FORWARD DISTANCE
RIGHT 90
SPI DISTANCE + 5
END
```

The command SPI 100 invokes a SPI agent and gives it the input message 100. The SPI agent then issues three commands. The first is just like the first command of the STEP agent: Tell the Turtle to go forward 100 units. The second tells the Turtle to turn right. Again there is nothing new. But the third does something extraordinary. This command is SPI 105. What is its effect? It tells the Turtle to go forward 105 units, tells the Turtle to turn right 90, and then issues the command SPI 110. Thus we have a trick called "recursion" for setting up a never-ending process whose initial steps are shown in figure 9.

Of all ideas I have introduced to children, recursion stands out as the one idea that is particularly able to evoke an excited response. I

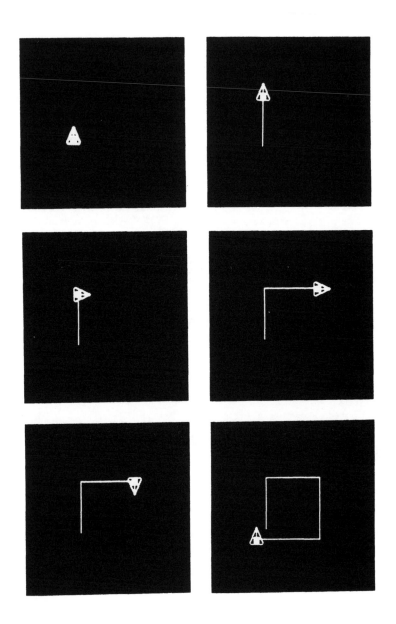

Figure 9a

Figure 9b

think this is partly because the idea of going on forever touches on every child's fantasies and partly because recursion itself has roots in popular culture. For example, there is the recursion riddle: If you have two wishes what is the second? (Two more wishes.) And there is the evocative picture of a label with a picture of itself. By opening the rich opportunities of playing with infinity the cluster of ideas represented by the SPI procedure puts a child in touch with something of what it is like to be a mathematician. Another aspect of living a mathematical experience is illustrated by figure 9b where we see how a curious mathematical phenomenon can be explored by varying the angle in the SPI procedure. Angles close to 90 produce a surprising emergent phenomenon: The arms of the galaxy like twisted squirals were not actually programmed into the procedure. They come as a shock and often motivate long explorations in which numerical and geometric thinking intertwines with aesthetics.

In the LOGO environment new ideas are often acquired as a means of satisfying a personal need to do something one could not do before. In a traditional school setting, the beginning student encounters the notion of variable in little problems such as:

$$5 + X = 8. \text{ What is X?}$$

Few children see this as a personally relevant problem, and even fewer experience the method of solution as a source of power. They are right. In the context of their lives, they can't do much with it. In the LOGO encounter, the situation is very much different. Here the child has a personal need: To make a spiral. In this context the idea of a variable is a source of personal power, power to do something desired but inaccessible without this idea. Of course, many children who encounter the notion of variable in a traditional setting do learn to use it effectively. But it seldom conveys a sense of "mathpower," not even to the mathematically best and brightest. And this is the point of greatest contrast between an encounter with the idea of variables in the traditional school and in the LOGO environment. In LOGO, the concept empowers the child, and the child experiences what it is like for mathematics to enable whole cultures to do what no one could do before.

If the use of a variable to make a spiral were introduced as an isolated example to "illustrate" the "concept of mathpower," it would have only a haphazard chance of connecting with a few children (as gears connected with me). But in Turtle geometry it is not an isolated example. It is typical of how all mathematical knowledge is encountered. *Mathpower*, one might say, *becomes a way of life.* The sense of power is not only associated with immediately applicable methods such as the use of angular measure of variables, but also with such concepts as "theorem" or "proof" or "heuristic" or "problem-solving method." In using these concepts, the child is developing ways to talk *about* mathematics. And it is to this development of mathematical articulateness we now turn.

Consider a child who has already made the Turtle draw a square and a circle and would now like to draw a triangle with all three sides equal to 100 Turtle steps. The form of the program might be:

```
TO TRIANGLE
REPEAT 3
   FORWARD 100
   RIGHT SOMETHING
END
```

But for the Turtle to draw the figure, the child needs to tell it more. What is the quantity we called SOMETHING? For the square we instructed the Turtle to turn 90 degrees at each vertex, so that the square program was:

```
TO SQUARE
REPEAT 4
   FORWARD 100
   RIGHT 90
END
```

Now we can see how Polya's precept, "find similarities," and Turtle geometry's procedural principle, "play Turtle," can work together. *What is the same* in the square and the triangle? If we play Turtle and "pace out" the trip that we want the Turtle to take, we notice that in both cases we start and end at the same point and facing the same direction. That is, we end in the state in which we started. And in between we did one complete turn. *What is differ-*

ent in the two cases is whether our turning was done "in three goes" or "in four goes." The mathematical content of this idea is as powerful as it is simple. Priority goes to the notion of the total trip—how much do you turn all the way around?

The amazing fact is that all total trips turn the same amount, 360 degrees. The four 90 degrees of the square make 360 degrees, and since all the turning happens at the corner the three turns in a triangle must each be 360 degrees divided by three. So the quantity we called SOMETHING is actually 120 degrees. This is the proposition of "The Total Turtle Trip Theorem."

> If a Turtle takes a trip around the boundary of any area and ends up in the state in which it started, then the sum of all turns will be 360 degrees.[5]

Part and parcel of understanding this is learning a method of using it to solve a well-defined class of problems. Thus the child's encounter with this theorem is different in several ways from memorizing its Euclidean counterpart: "The sum of the internal angles of a triangle is 180 degrees."

First (at least in the context of LOGO computers), the Total Turtle Trip Theorem is more *powerful*: The child can actually use it. Second, it is more *general*: It applies to squares and curves as well as to triangles. Third, it is more *intelligible*: Its proof is easy to grasp. And it is more personal: You can "walk it through," and it is a model for the general habit of relating mathematics to personal knowledge.

We have seen children use the Total Turtle Trip Theorem to draw an equilateral triangle. But what is exciting is to watch how the theorem can accompany them from such simple projects to far more advanced ones—the flowers in the boxes that are reproduced in the center of the book show a project a little way along this path. For what is important when we give children a theorem to use is not that they should memorize it. What matters most is that by growing up with a few very powerful theorems one comes to appreciate how certain ideas can be used as tools to think with over a lifetime. One learns to enjoy and to respect the power of powerful ideas. One learns that the most powerful idea of all is the idea of powerful ideas.

What follows is a hypothetical conversation between two children who are working and playing with the computer. These and other experiments can happen every day—and they do.

A PLAN

——Let's make the computer draw a flower like this.

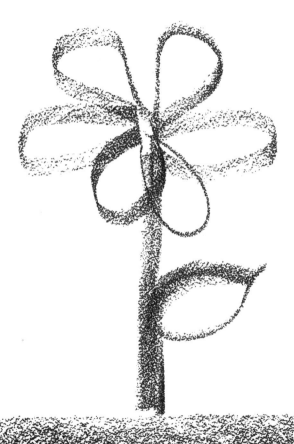

FIND RESOURCES

——Do you have any programs we could use?
——Yes, there's that quarter circle thing I made last week.
——Show me.

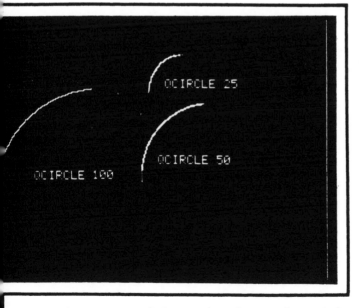

——It draws quarter circles starting wherever the turtle is.
——It needs an input to tell it how big.

TRY SOMETHING

——Let's make a petal by putting two QCIRCLES togethe[r]
——OK. What size?
——How about 50?

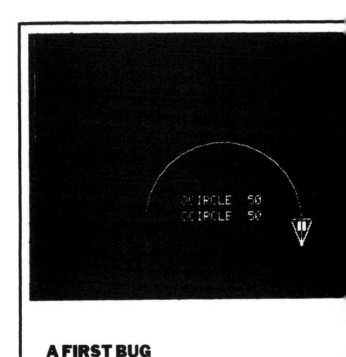

A FIRST BUG

——It didn't work.
——Of course! Two QCIRCLES make a semicircle.

FIX THE BUG

——We have to turn the Turtle between QCIRCLES.
——Try 120°.
——OK, that worked for triangles.
——And let's hide the Turtle by typing HIDETURTLE.

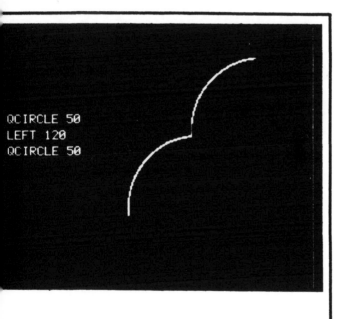

IT'S A BIRD!

——What's going on?
——Try a right turn.

——Why don't we just stick with the bird? We could m[a]
a flock.
——You do that. I want my flower.
——We could do the flower, then the flock.

```
DCIRCLE 50
RIGHT 120
DCIRCLE 50
```

IT'S A FISH!

——The right turn is better.
——But we don't know how much to turn.
——We could try some more numbers.
——Or we could try some mathematics.

MATH TO THE RESCUE

——Do you know about the Total Turtle Trip Theorem? You think about the Turtle going all around the petal and add up all the turns. 360°.

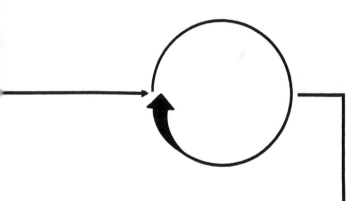

——All around is 360.

——Each QCIRCLE turns it 90. That makes 180 for two QCIRCLES.

——360 altogether. Take away 180 for the QCIRCLES. That leaves 180 for the pointy parts. 90 each.

——So we should do RIGHT 90 at each point.

——Let's try.

```
TO PETAL

QCIRCLE 50
RIGHT 90
QCIRCLE 50
RIGHT 90
END
```

——Four make a flower.

```
TO FLOWER

PETAL
RIGHT 90
PETAL
RIGHT 90
PETAL
RIGHT 90
PETAL
RIGHT 90
END
```

——That's more like a propeller.

——So try ten.

——If we let PETAL have an input we can make big or small flowers.
——That's easy. Just do TO PETAL SIZE QCIRCLE SIZE, and so on.
——But I bet we'd get bugs if we try that. Let's try plain 25 first.
——Then we can make a superprocedure to draw a plant.

BUILDING UP

```
TO PLANT
NEWFLOWER
BACK 50
PETAL
BACK 50
END
```

A BUILDING BLOCK

——Typing all that ten times hurts my fingers.
——We can use REPEAT.

```
TO NEWFLOWER
REPEAT 10
    PETAL
    RIGHT 360/10
END
```

——There it is!
——But it's too big.
——All we have to do is change the 50 in PETAL.
  Make it 25.

—I have a great procedure for putting several together. It's called SLIDE.
You just go, PLANT SLIDE PLANT SLIDE PLANT SLIDE.

TO SLIDE DISTANCE
PEN UP
RIGHT 90
FORWARD DISTANCE
LEFT 90
PEN DOWN   END

TRYING THE NEW TOOL

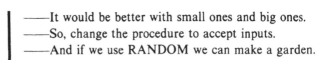

——It would be better with small ones and big ones.
——So, change the procedure to accept inputs.
——And if we use RANDOM we can make a garden.

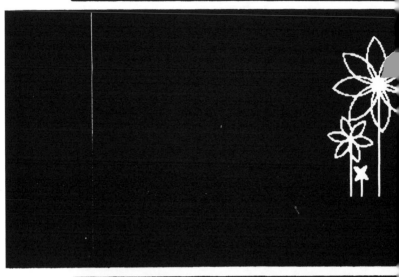

——My next project is a flock of birds.

——Maybe we'll put the birds and flowers together.

——Maybe.

——Make a flock by doing BIRD SLIDE BIRD SLIDE.
——I want six birds, and I'm going to use REPEAT.

——That's funny. I wanted 6 birds all the same way up.
——But it's neat. If we debug it, we should keep a copy
   like this.

BIRD
SLIDE
BIRD

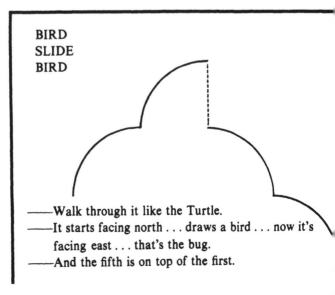

——Walk through it like the Turtle.
——It starts facing north . . . draws a bird . . . now it's
   facing east . . . that's the bug.
——And the fifth is on top of the first.

——If you want to fix the bug, bring the Turtle around to face north after doing the bird.
——And let's make them smaller.

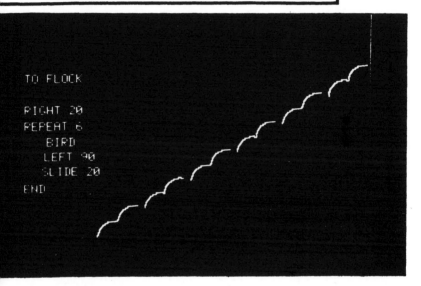

```
TO FLOCK

RIGHT 20
REPEAT 6
    BIRD
    LEFT 90
    SLIDE 20
END
```

——Here's the flock.

THE END

AND . .

——It's not finished. Let's give the flock inputs and put several together.

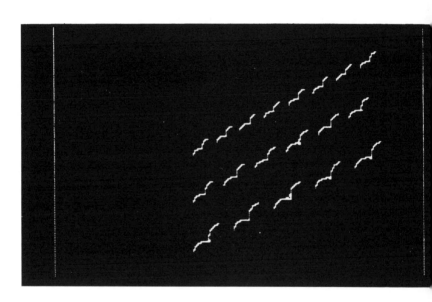

——How can we make them fly?
——I found something neat. In BIRD use SPIN instead of RIGHT . . . it's got bugs, but it is pretty.

. . . A BEGINNING

The next phase of the project will produce the most spectacular effects as the birds go into motion. But the printed page cannot capture either the product or the process: the serendipitous discoveries, the bugs, and the mathematical insights all require movement to be appreciated. Reflecting on what you are missing leads me to another description of something new the computer offers a child: the opportunity to draw in motion, indeed to doodle and even to scribble with movement as well as with lines. Perhaps they will be learning, as they do so, to think more dynamically.

Chapter 4

Languages for Computers and for People

The Centipede was happy quite
Until the toad in fun
Said, Pray which leg comes after which?
This wrought her mind to such a pitch
She lay distracted in a ditch
Considering how to run

—Anonymous

THE CENTIPEDE STORY is disturbing. We usually like to think that thinking and understanding are, by definition, good things to do, and that, in particular, they are useful in learning. But the centipede came to grief by thinking about her own actions. Would the same thing happen to us? Does this mean we should give up thinking about ourselves? In fact, in our "rational" culture, the notion that thinking impedes action, even that thinking impedes learning, is quite prevalent. It is our usual way of talking about learning to ride a bicycle: "Keep trying—one day you'll just 'get it' " is standard parental advice to children struggling with the two-wheeler.

Many philosophers have developed the idea that some knowledge cannot be described in words or grasped by conscious thought. The idea was brought into recent curriculum reforms by advocates of active learning and given theoretical support by J. S. Bruner's[1] influential classification of ways of knowing: Some knowledge is represented as action, some as image, and only the third category as symbols. Bruner has asserted that "words and diagrams" are "impotent" to represent certain kinds of knowledge which are only representable as action. In this chapter I try to develop a more flexible perspective on these problems.

My perspective is more flexible because it rejects the idea of the dichotomy verbalizable versus nonverbalizable. No knowledge is entirely reducible to words, and no knowledge is entirely ineffable. My perspective is more flexible also in recognizing a historical dimension: An important component in the history of knowledge is the development of techniques that increase the potency of "words and diagrams." What is true historically is also true for the individual: An important part of becoming a good learner is learning how to push out the frontier of what we can express with words. From this point of view the question about the bicycle is not whether or not one can "tell" someone "in full" how to ride but rather what can be done to improve our ability to communicate with others (and with ourselves in internal dialogues) just enough to make a difference to learning to ride. The central theme of this chapter is the development of descriptive languages for talking about learning. We shall focus particular attention on one of the kinds of learning that many people believe to be best done by "just doing it"—the learning of physical skills. Our approach to this is the exact opposite of the way schools treat "physical education"—as a nonintellectual subject. Our strategy is to make visible even to children the fact that learning a physical skill has much in common with building a scientific theory.

With this realization comes many benefits. First, I know from work in the LOGO laboratory that it means more effective learning of physical skills.[2] Without this direct benefit, seeking to "motivate" a scientific idea by drawing an analogy with a physical activity could easily degenerate into another example of "teacher's double talk." But if we can find an honest place for scientific thinking in activities that the child feels are important and personal, we shall open the doors to a more coherent, syntonic pattern of learning.

In this chapter I show that this can be done and suggest that relating science to physical skills can do much more for learning science than providing what educators like to call a "motivation." It can potentially place children in a position of feeling some identification with scientists through knowing that scientists use formal descriptive languages and knowing that they too can use such languages as tools for learning physical skills—juggling for example. The idea is to give children a way of thinking of themselves as "doing science" when they are doing something pleasurable with their bodies. If children could see Descartes's invention of coordinate geometry as something not totally alien to their own experiences of daily life, this could not only make Descartes more meaningful but, at the same time, help the children come to see themselves as more meaningful.

Let us look a bit more closely at what our culture thinks about learning physical skills. It is no more consistent regarding this than it is regarding the mathematics of more "abstract" subjects we discussed earlier. Although the popular wisdom and much of educational psychology may agree that learning physical skills is a domain where "conscious" thinking doesn't help, people who make sports their livelihood don't always agree. Some of the most successful coaches put great effort into analyzing and verbalizing the movements that must be learned and perfected. One sportswriter, Timothy Gallwey, has turned popular sensitivity to this contradiction into publishing success. In his book *Inner Tennis* he offers some suggestions for a way out of the dilemma. Gallwey encourages the learner to think of himself as made up of two selves: an analytic, verbal self and a more holistic, intuitive one. It is appropriate, he argues, that now one and now the other of these two selves should be in control; in fact, an important part of the learning process is teaching each "self" to know when to take over and when to leave it to the other.

Gallwey's description of the negotiation and transactions that go with successful learning is unusual in educational circles. In the choice between analytic and holistic modes of thinking, he gives control to the learner. This is very different from what usually happens in curriculum design for schools. Curriculum reformers are often concerned about the choice between verbal and nonverbal, experimental learning. But their strategy is usually to make the choice from above and build it into the curriculum. Gallwey's strategy is to help learners learn how to make the choice for themselves, a perspective that is in line with the vision already suggested of the child as epistemologist, where the child is encouraged to become expert in recognizing and choosing among varying styles of thought.

Taking Timothy Gallwey as an example is not an endorsement of everything he says. Most of his ideas strike me as problematic. But I think he is quite right in recognizing that people need more structured ways to talk and think about the learning of skills. Contemporary language is not sufficiently rich in this domain.

In a computer-rich world, computer languages that simultaneously provide a means of control over the computer and offer new and powerful descriptive languages for thinking will undoubtedly be carried into the general culture. They will have a particular effect on our language for describing ourselves and our learning. To some extent this has already occurred. It is not uncommon for people with no knowledge of computers to use such concepts as "input," "output," and "feedback" to describe their own mental processes. We shall give an example of this process by showing how programming concepts can be used as a conceptual framework for learning a particular physical skill, namely, juggling. Thus we look at programming as a source of descriptive devices, that is to say as a means of strengthening language.

Many scientific and mathematical advances have served a similar linguistic function by giving us words and concepts to describe what had previously seemed too amorphous for systematic thought. One of the most striking examples of the power of descriptive language is the genesis of analytic geometry, which played so decisive a role in the development of modern science.

Legend has it that Descartes invented analytic geometry while lying in bed late one morning observing a fly on the ceiling. We can imagine what his thinking might have been. The fly moving hither and thither traced a path as real as the circles and ellipses of Euclidean mathematics, but one that defied description in Euclidean language. Descartes then saw a way to grasp it: At each moment the fly's position could be described by saying how far it was from the walls. Points in space could be described by pairs of numbers; a path could be described by an equation or relationship that holds true for those number pairs whose points lie on the path. The potency of symbols took a leap forward when Descartes realized how to use an algebraic language to talk about space, and a spatial language to talk about algebraic phenomena. Descartes's method of coordinate geometry born from this insight provided tools that science has since used to describe the paths of flies and planets and the "paths" of the more abstract objects, the stuff of pure mathematics.

Descartes's breakthrough has much in common with the experience of the child in the Turtle circle episode. The child, we recall, was explicitly looking for a way to describe the process of walking in a circle. In LOGO this description takes a very simple form, and the child has to invent only the description. Descartes had to do more. But in both cases the reward is the ability to describe analytically something that until then was known in a global, perceptual-kinesthetic way. Neither the child nor Descartes suffered the fate of the centipede: Both could walk in circles as well after knowing how to describe their movements analytically as before.

But Descartes's formalism, powerful as it is for describing processes in the world of physics, is not what is needed for describing processes in the world of physical skills.

Using calculus to describe juggling or how a centipede walks would indeed be confusing. Attempts to use such descriptions in learning physical skills very likely *would* leave the learner lying with feverish mind in the nearest ditch. This mode of formal description is not matched to this task. But other formalisms are.

The field of education research has not worked in the direction of developing such formalisms. But another research community, that of computer scientists, has had (for its own reasons) to work on the problem of descriptive languages and has thereby become an unexpected resource for educational innovation. Computers are called upon to do many things, and getting a computer to do something requires that the underlying process be described, on some level, with enough precision to be carried out by the machine. Thus computer scientists have devoted much of their talent and energy to developing powerful descriptive formalisms. One might even say that computer science is wrongly so called: Most of it is not the science of computers, but the science of descriptions and descriptive languages. Some of the descriptive formalisms produced by computer science are exactly what are needed to get a handle on the process of learning a physical skill. Here we demonstrate the point by choosing one important set of ideas from programming: the concept of structured programming, which we shall illustrate by the learning experience of a fifth grader in a LOGO environment.

Keith had set himself the goal of making the computer draw a stick figure as in the box GOAL (see Figure 10a).

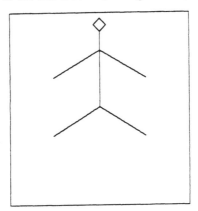

Figure 10a
Goal

His plan was to start with one foot and draw the Turtle strokes illustrated in the box SEQUENCE. In doing so he is using an image familiar in his precomputational culture, where he has learned to do connect-the-dots drawing and to describe his activities in a step-by-step way. So it is perfectly natural for him to adopt this method here. The task seemed simple if somewhat tedious. He wrote (Figure 10b):

```
TO MAN
FORWARD 300
RIGHT 120
FORWARD 300
RIGHT 180
FORWARD 300
LEFT 120
FORWARD 300
LEFT 120
FORWARD 300
 RIGHT 180
 FORWARD 300
 RIGHT 120
 FORWARD 300
 RIGHT 180
 FORWARD 300
 LEFT 120
 FORWARD 150
 LEFT 45
 FORWARD 50
 RIGHT 90
 FORWARD 50
 RIGHT 90
 FORWARD 50
 RIGHT 90
 FORWARD 50
 END
```

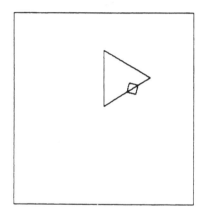

Figure 10b

Figure 10c
Bugged Man

What appeared on the screen was the totally unexpected drawing of the BUGGED MAN (see Figure 10c). What went wrong?

Keith was prepared for surprises of this sort. As mentioned earlier, one of the mainstays of the LOGO environment is the cluster of concepts related to "bugs" and "debugging." One does not expect anything to work at the first try. One does not judge by standards like "right—you get a good grade" and "wrong—you get a bad grade." Rather one asks the question: "How can I fix it?" and to fix it one has first to understand what happened in its own terms.

101

MINDSTORMS

Only then can we make it happen on our terms. But in this situation Keith was unable to figure out what had happened. He had written his program in such a way that it was extremely difficult to pinpoint his error. Where was the bug in his program? What error could cause such a wild transformation of what he had intended?

In order to understand his predicament we contrast his program with a different strategy of programming known as "structured programming." Our aim is to subdivide the program into natural parts so that we can debug programs for each part separately. In Keith's long, featureless set of instructions it is hard to see and trap a bug. By working with small parts, however, bugs can be confined and more easily trapped, figured out. In this case a natural subdivision is to make a program to draw a V-shaped entity to use for arms and legs and another to draw a square for the head. Once these "subprocedures" have been written and tested, it is extremely easy to write the "superprocedure" to draw the stick figure itself. We can write an extremely simple program to draw the stick figure:

```
TO MAN
VEE
FORWARD 50
VEE
FORWARD 25
HEAD
END
```

This procedure is simple enough to grasp as a whole. But of course it achieves its simplicity only by making the assumption that the commands VEE and HEAD are understood by the computer. If they are not, the next step must be to define VEE and HEAD. We can do this in the same style of always working with a procedure we can understand as a whole. For example:

```
TO VEE
RIGHT 120
LINE 50
RIGHT 120
LINE 50
RIGHT 120
END
```

(In this program we assume we have defined the command LINE, which causes the Turtle to go forward and come back.)

To make this work we next define LINE:

```
TO LINE :DISTANCE
FORWARD :DISTANCE
BACK :DISTANCE
END
```

Since the last procedure uses only innate LOGO commands, it will work without further definitions. To complete MAN we define HEAD by:

```
TO HEAD
RIGHT 45
SQUARE 20
END
```

Robert, a seventh grader, expressed his conversion to this style of programming by exclaiming: "See, all my procedures are mind-sized bites." Robert amplified the metaphor by comments such as: "I used to get mixed up by my programs. Now I don't bite off more than I can chew." He had met a powerful idea: It is possible to build a large intellectual system without ever making a step that cannot be comprehended. And building with a hierarchical structure makes it possible to grasp the system as a whole, that is to say, to see the system as "viewed from the top."

Keith, the author of the nonstructured MAN program, had been exposed to the idea of using subprocedures but had previously resisted it. The "straight-line" form of program corresponded more closely to his familiar ways of doing things. He had experienced no compelling need for structured programming until the day he could not debug his MAN program. In LOGO environments we have seen this happen time and again. When a child in this predicament asks what to do, it is usually sufficient to say: "You *know* what to do!" And often the child will say, sometimes triumphantly, sometimes sheepishly: "I guess I should turn it into subprocedures?" The "right way" was not imposed on Keith; the computer gave him enough flexibility and power so that his exploration could be genuine and his own.

These two styles of approaching the planning and working out of a project are pervasive. They can be seen by observing styles of learning "physical" as well as "intellectual" skills. Consider, for example, the case of two fifth graders who learned both programming and physical skills in our children's learning laboratory.

Michael is strong, athletic, a "tough kid" in his own eyes. Paul is more introverted, studious, slightly built. Michael does poorly at school and Paul does well, so when Paul got on faster in work with the computer, moving quickly into quite complex structured programming procedures, neither one was surprised. After several weeks Michael was still able to write programs only in the straight-line style. There was no doubt that he possessed all the necessary concepts to write more elaborate programs, but he was held back by a classical and powerful resistance to using subprocedures.

At this time both began to work on stilt walking. Michael's strategy was to fix in his mind a model of stilt walking in sequential form: "Foot on the bar, raise yourself up, foot on the other bar, first foot forward..." When attempting to do it led to a rapid crash, he would bravely start again and again and again, confident that he would eventually succeed, which in fact he did. But, to the surprise of both of them, Paul got there first.

Paul's strategy was different. He began in the same way but when he found that he was not making progress he tried to isolate and correct part of the process that was causing trouble: "the bug." When you step forward you tend to leave the stilt behind. This bug, once identified, is not hard to eradicate. One trick for doing so is to think of taking the step with the stilt rather than with the foot and let the stilt "carry" the foot. This is done by lifting the stilt with the arm against the foot. The analogy with his approach to programming was so apparent to Paul that this might have been a case of "transfer" from the programming work to learning this physical skill.

Actually, it is more likely that both situations drew on long-standing features of his general cognitive style. But the experience with LOGO did enable Paul to articulate these aspects of his style. The relation between programming and stilt walking was even clearer in Michael's case. It was only through this analogy that he came to recognize explicitly the difference between his and Paul's style of going about the stilt walking! In other words the experience of programming helped both boys obtain a better grasp of their own actions, a more articulated sense of themselves.

The generality of the idea of structured programming as a mathetic principle, that is to say an aid to learning, will become more apparent through the next example, which describes the process involved in learning another physical skill—juggling. We do not choose it at random. The Turtle circle was a good carrier for learning mathematics "with one's body." Juggling turns out to be an equally good carrier for learning a body skill "with mathematics." Of course, the picture is more complicated and also more interesting because in both cases the process works in both directions, from computational metaphor to body language and back again. In passing through an experience of Turtle geometry, children sharpen their sense of their bodies and their physical movements as well as their understanding of formal geometry. And theoretical ideas about structured programs, when couched in juggling terms—real body terms—take on new concreteness and power. In both cases, knowledge takes on the quality we have characterized as syntonic.

There are many different kinds of juggling. When most people think of juggling, they are thinking about a procedure that is called "showers juggling." In showers juggling balls move one behind the other in a "circle" passing from left to right at the top and from right to left at the bottom (or vice versa). This takes two kinds of throws: a short, low throw to get the balls from one hand to the other at the bottom of the "circle" (near the hands), and a long, high throw to get the balls to go around the top of the circle. (See Figure 11.)

Cascade juggling has a simpler structure. There is no bottom of the circle; balls travel in both directions over the upper arc. There

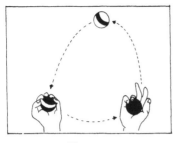 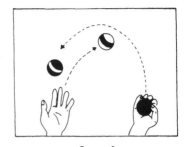

Showers   **Cascade**

Figure 11
Two Forms of Juggling

is only one kind of toss: a long and high one. (See Figure 11.) Its simplicity makes it a better route into juggling as well as a better example for our argument. Our guiding question is this: Will someone who wishes to learn cascade juggling be helped or hindered by a verbal, analytic description of how to do it? The answer is: It all depends. It depends on what materials the learner has for making analytic descriptions. We use cascade juggling to show how good computational models can help construct "people procedures" that improve performance of skills and how reflection on those people procedures can help us learn to program and to do mathematics. But, of course, *some* verbal descriptions will confuse more than they will help. Consider, for example, the description:

1. Start with balls 1 and 2 in the left hand and ball 3 in the right.
2. Throw ball 1 in a high parabola to the right hand.
3. When ball 1 is at the vertex throw ball 3 over to the left hand in a similar high parabola, but take care to toss ball 3 under the trajectory of ball 1.
4. When ball 1 arrives at the right hand and ball 3 is at the vertex, catch ball 1 and throw ball 2 in a trajectory under that of ball 3, and so on.

This description is basically a brute-force straight-line program. It is not a useful description for the purpose of learning. People outside the computer culture might say it is too much like a computer program, "just one instruction after another." It is like certain programs, for example Keith's first MAN program. But we have seen that stringing instructions together without good internal structure is not a good model for computer programming either, and we shall see that the techniques of structured programming that *are* good for writing programs are also good as mathetic descriptions of juggling.

Our goal is to create a people procedure: TO JUGGLE. As a first step toward defining this procedure we identify and name subprocedures analogous in their role to the subprocedures Keith used in drawing his stick figure (TO VEE, TO HEAD, TO LINE). In the case of juggling, a natural pair of subprocedures is what we call TOSSRIGHT and TOSSLEFT. Just as the command VEE was defined functionally by the fact that it causes the computer to place a certain V-shaped figure on the screen, the command TOSS-LEFT given to our apprentice juggler should "cause" him to throw a ball, which we assume he is holding in his left hand, over to the right hand.

But there is an important difference between programming TO MAN and programming TO JUGGLE. The programmer of TO MAN need not worry about timing, but in setting up the procedure for juggling we *must* worry about it. The juggler must perform the actions TOSSRIGHT and TOSSLEFT at appropriate moments in a cycle, and the two actions will have to overlap in time. Since we have chosen to include the catching phase in the same subprocedure as the throwing phase, the procedure TOSSRIGHT is meant to include catching the ball when it comes over to the left hand. Similarly, TOSSLEFT is a command to throw a ball from the left hand over to the right and catch it when it arrives.[3]

Since most people can perform these actions, we shall take TOSSLEFT and TOSSRIGHT as given and concentrate on how they can be combined to form a new procedure we shall call TO JUGGLE. Putting them together is different in one essential way from the combination of subprocedures TO VEE and TO HEAD to make the procedure TO MAN. TOSSLEFT might have to be initiated before the action initiated by the previous TOSSRIGHT is completed. In the language of computer science, this is expressed by saying that we are dealing with *parallel* processes as opposed to the strictly *serial* processes used in drawing the stick figure.

To describe the combination of the subprocedures we introduce a new element of programming: The concept of a "WHEN DE-

MON." This is illustrated by the instruction: WHEN HUNGRY EAT. In one version of LOGO this would mean: Whenever the condition called HUNGRY happens, carry out the action called EAT. The metaphor of a "demon" expresses the idea that the command creates an autonomous entity within the computer system, one that remains dormant until a certain kind of event happens, and then, like a demon, it pounces out to perform its action. The juggling act will use two such WHEN DEMONS.

Their definitions will be something like:

> WHEN something TOSSLEFT
> WHEN something TOSSRIGHT

To fill the blanks, the "somethings," we describe two conditions, or recognizable states of the system, that will trigger the tossing action.

At a key moment in the cycle the balls are disposed about like this (Figure 12):

Left Hand   Right Hand

Figure 12

But this diagram of the state of the system is incomplete since it fails to show in which direction the top ball is flying. To complete it we add arrows to indicate a direction (Figure 13a) and obtain two state descriptions (Figures 13b and 13c).

Figure 13a

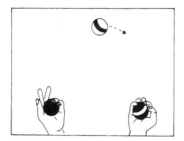

Figure 13b
TOPRIGHT: The ball is at the top and is moving to the right

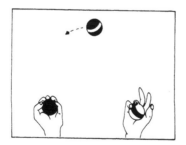

Figure 13c
TOPLEFT: The ball is at the top and is moving to the left

If we assume, reasonably, that the juggler can recognize these two situations, the following formalism should be self-explanatory:

                    TO KEEP JUGGLING
                    WHEN TOPRIGHT TOSSRIGHT
                    WHEN TOPLEFT TOSSLEFT

or even more simply:

                    TO KEEP JUGGLING
                    WHEN TOPX TOSSX

which declares that when the state TOPRIGHT occurs, the right hand should initiate a toss and when TOPLEFT occurs, the left hand should initiate a toss. A little thought will show that this is a complete description: The juggling process will continue in a self-perpetuating way since each toss creates a state of the system that triggers the next toss.

How can this model that turned juggling into a *people procedure* be applied as a teaching strategy? First, note that the model of juggling made several assumptions:

1. that the learner can perform TOSSRIGHT and TOSSLEFT
2. that she can recognize the trigger states TOPLEFT and TOPRIGHT
3. that she can combine these performance abilities according to the definitions of the procedure TO KEEP JUGGLING

Now, we translate our assumptions and our people procedure into a teaching strategy.

STEP 1: Verify that the learner *can* toss. Give her one ball, ask her to toss it over into the other hand. This usually happens smoothly, but we will see later that a minor refinement is often needed. The spontaneous procedure has a bug.

STEP 2: Verify that the learner can combine tosses. Try with two balls with instructions:

                    TO CROSS
                    TOSSLEFT
                    WHEN TOPRIGHT TOSSRIGHT
                    END

This is intended to exchange the balls between left and right hands. Although it appears to be a simple combination of TOSSLEFT and TOSSRIGHT, it usually does not work immediately.

STEP 3. Look for bugs in the toss procedures. Why doesn't TO CROSS work? Typically we find that the learner's ability to toss is not really as good as it seemed in step 1. The most common deviation or "bug" in the toss procedure is following the ball with the eyes in doing a toss. Since a person has only one pair of eyes, their engagement in the first toss makes the second toss nearly impossible and thus usually ends in disaster with the balls on the floor.

STEP 4. Debugging. Assuming that the bug was following the first ball with the eyes, we debug by returning our learner to tossing with one ball without following it with her eyes. Most learners find (to their amazement) that very little practice is needed to be able to perform a toss while fixing the eyes around the expected apex of the parabola made by the flying ball. When the single toss is debugged, the learner again tries to combine two tosses. Most often this now works, although there may still be another bug to eliminate.

STEP 5. Extension to three balls. Once the learner can smoothly execute the procedure we called CROSS, we go on to three balls. To do this begin with two balls in one hand and one in the other (Figure 14).

Ball 2 is tossed as if executing CROSS, ignoring ball 1. The TOSS-RIGHT in CROSS brings the three balls into a state that is ready for KEEP JUGGLING. The transition from CROSS to KEEP JUGGLING presents a little difficulty for some learners, but this is easily overcome. Most people can learn to juggle in less than half an hour by using this strategy.

Variants of this teaching strategy have been used by many LOGO teachers and studied in detail by one of them, Howard Austin, who took the analysis of juggling as the topic of his Ph.D. thesis. There is no doubt that the strategy is very effective and little doubt as to the cause: The use of programming concepts as a descriptive language facilitates debugging.

In our description of drawing a stick figure and of learning to juggle, a central theme was how debugging is facilitated by the use of an appropriate description of a complex process. In both cases

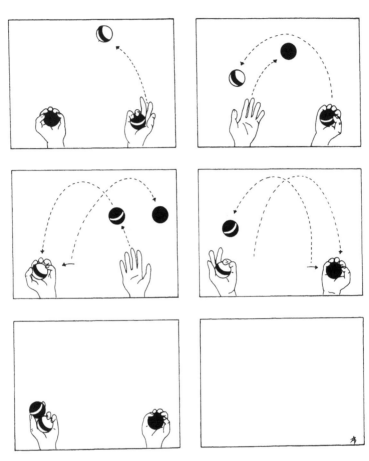

Figure 14
Cascade Juggling

the description reflected a representation of the process in modular form, that is to say broken up into natural, functional units, and catching the bug was helped by containing it within as narrow a set of boundaries as possible. The worst conditions for debugging are created when several bugs are present simultaneously. The debugging process is especially effective if the modules are small enough for it to be unlikely that any one contains more than one bug.

The difficulties produced by multiple bugs are well illustrated by what happens when beginners try to learn juggling by the brute-force approach. In fact (just as Michael learned to walk on stilts) they often succeed, usually after hours of frustrating attempts to keep three balls in the air without yet being able to cross two. But it takes them a long time to learn. When Howard Austin looked closely at the actions of the learner, he saw the same bugs that we described in our rational strategy approach, for example, the eye-following bug. In the course of very many repetitions, so-called "trial and error learning" will shape a behavior that works. By sheer chance, the eyes will happen to move a little less on one toss. Like other animals, human beings have learning mechanisms that are capable of picking up on such events and increasing the probability that they will happen again. Eventually, the bugs are ironed out and the subject learns to juggle. People are capable of learning like rats in mazes. But the process is slow and primitive. We can learn more, and more quickly, by taking conscious control of the learning process, articulating and analyzing our behavior.

The fact that computational procedures enhance learning does not mean that all repetitive processes can be magically removed from learning or that the time needed to learn juggling can be reduced to almost nothing. It always takes time to trap and eliminate bugs. It always takes time to learn necessary component skills. What can be eliminated are wasteful and inefficient methods. Learning enough juggling skill to keep three balls going takes many hours when the learner follows a poor learning strategy. When a good one is adopted the time is greatly reduced, often to as little as twenty or thirty minutes.

Children often develop a "resistance" to debugging analogous to the resistance we have seen to subprocedurizing. I have seen this in many childrens' first sessions in a LOGO environment. The child plans to make the Turtle draw a certain figure, such as a house or stick man. A program is quickly written and tried. It doesn't work. Instead of being debugged, it is erased. Sometimes the whole project is abandoned. Sometimes the child tries again and again and again with admirable persistence but always starting from scratch in an apparent attempt to do the thing "correctly" in one shot. The child might fail or might succeed in making the computer draw the picture. But this child has not yet succeeded in acquiring the strategy of debugging.

It is easy to empathize. The ethic of school has rubbed off too well. What we see as a good program with a small bug, the child sees as "wrong," "bad," "a mistake." School teaches that errors are bad; the last thing one wants to do is to pore over them, dwell on them, or think about them. The child is glad to take advantage of the computer's ability to erase it all without any trace for anyone to see. The debugging philosophy suggests an opposite attitude. Errors benefit us because they lead us to study what happened, to understand what went wrong, and, through understanding, to fix it. Experience with computer programming leads children more effectively than any other activity to "believe in" debugging.

Contact with the LOGO environment gradually undermines long-standing resistances to debugging and subprocedurizing. Some people who observe the childrens' growing tolerance for their "errors" attribute the change of attitude to the LOGO teachers who are matter-of-fact and uncritical in the presence of programs the child sees as "wrong." I think that there is something more fundamental going on. In the LOGO environment, children learn that the teacher too is a learner, and that *everyone* learns from mistakes.

A group of twelve fifth graders had had several hours a week of LOGO experience since the beginning of the term in September. Early in December the group decided on a collective project. A mechanical Turtle would be programmed to write "Merry Christmas" on huge paper banners that would be strung in the school corridors. An ideal project. The letters of the alphabet were divided up among members of the group. Each would write programs for two or three letters, for decorative drawings, and for whole messages, using the letter programs as subprocedures.

But snowstorms and other disruptions delayed the work; and when the last week of school arrived the banners had not yet been made. The instructor in charge of the group decided to break a general rule and to do some of the programming herself. She worked at home without a Turtle so when she came in the next morning she had a collection of un-debugged programs. She and the children would debug them together. The instructor and a child were on the floor watching a Turtle drawing what was meant to be a letter R, but the sloping stroke was misplaced. Where was the bug? As they puzzled together the child had a revelation: "Do you mean," he said, "that you really don't know how to fix it?" The child did not yet know how to say it, but what had been revealed to him was that he and the teacher had been engaged together in a research project. The incident is poignant. It speaks of all the times this child entered into teachers' games of "let's do that together" all the while knowing that the collaboration was a fiction. Discovery cannot be a setup; invention cannot be scheduled.

In traditional schoolrooms, teachers do try to work collaboratively with children, but usually the material itself does not spontaneously generate research problems. Can an adult and a child genuinely collaborate on elementary school arithmetic? A very important feature of work with computers is that the teacher and the learner can be engaged in a real intellectual collaboration; together they can try to get the computer to do this or that and understand what it actually does. New situations that neither teacher nor learner has seen before come up frequently and so the teacher does not have to pretend not to know. Sharing the problem and the experience of solving it allows a child to learn from an adult not "by doing what teacher says" but "by doing what teacher does." And one of the things that the teacher does is pursue a problem until it is completely understood. The LOGO environment is special because it provides numerous problems that elementary schoolchildren can understand with a kind of completeness that is rare in ordinary life. To appreciate the point more fully it may be useful to rethink the simple examples of debugging discussed earlier.

We have discussed the program:

```
TO HOUSE
SQUARE
TRIANGLE
END

TO SQUARE
REPEAT 4
    FORWARD 100
    RIGHT 90
END

TO TRIANGLE
REPEAT 3
    FORWARD 100
    RIGHT 120
END
```

But this program contains a bug and draws the triangle inside the square instead of on it. Why? It might seem mysterious at first to a child. But you can figure out "why the Turtle did that dumb thing" by following through on a already well-known piece of heuristic advice: Play Turtle. Do it yourself but pretend to be as dumb as the Turtle. Finding out why the Turtle did it almost immediately suggests a way to fix it. For example, some say: "The Turtle turned *into* the square because TRIANGLE says RIGHT TURN." A cure (one of several equally simple ones) is inherent in this diagnosis: Make a triangle procedure with left turns.

Similarly an adult who thought he could make the Turtle draw a triangle by REPEAT [FORWARD 100 RIGHT TURN 60] would be astonished to see a hexagon appear. But it is possible to "get into" the program and see why this happens. Moreover, it is possible to introspect and see how the bug came from a very superficial understanding of the most common statement of Euclid's triangle theorem: "The sum of the angles of a triangle is 180 degrees."

A child (and, indeed, perhaps most adults) lives in a world in which everything is only partially understood: well enough perhaps, but never completely. For many, understanding the Turtle's action so completely that there is nothing more to say about it is a rare, possibly unique, experience. For some it is an exhilarating one: We can see this by the children's eagerness to explain what they have understood. For all it is a better model of the crispness of analytic knowledge than most people ever encounter.

The reader might object that far from understanding the Turtle "fully" a child programmer hardly understands at all the complex mechanisms at work behind the scenes whenever a Turtle carries out a LOGO command. Are we in fact in danger of mystifying children by placing them in an environment of sophisticated technology whose complexities are only partially understood by advanced computer scientists?

These concerns bring us back full circle to the issues with which this chapter began. For example, I proposed a description of juggling in the form of a simple program. But the same concern arises: Does the description in procedural language grasp the essence of the process of juggling or does it mystify by covering over the complexities of the juggling?

These questions are very general and touch on fundamental issues of scientific method. Newton "understood" the universe by reducing whole planets to points that move according to a fixed set of laws of motion. Is this grasping the essence of the real world or hiding its complexities? Part of what it means to be able to think like a scientist is to have an intuitive understanding of these epistemological issues and I believe that working with Turtles can give children an opportunity to get to know them.

It is in fact easy for children to understand how the Turtle defines a self-contained world in which certain questions are relevant and others are not. The next chapter discusses how this idea can be developed by constructing many such "microworlds," each with its own set of assumptions and constraints. Children get to know what it is like to explore the properties of a chosen microworld undisturbed by extraneous questions. In doing so they learn to transfer habits of exploration from their personal lives to the formal domain of scientific theory construction.

The internal intelligibility of computer worlds offers children the opportunity to carry out projects of greater complexity than is usually possible in the physical world. Many children imagine complex structures they might build with an erector set or fantasize about organizing their friends into complex enterprises. But when they try to realize such projects, they too soon run into the unintelligible limitations of matter and people. Because computer programs can in principle be made to behave exactly as they are intended to, they can be combined more safely into complex systems. Thus, children are able to acquire a feel for complexity.

Modern science and engineering have created the opportunity for achieving projects of a degree of complexity scarcely imaginable until recently. But science teaches us the power of simplicity as well and I end the chapter with what I find to be a moving story of a child who learned something of this in a particularly simple but personally important experience.

Deborah, a sixth grader who had problems with school learning, was introduced to the world of screen Turtles by being shown how they could obey the commands FORWARD, LEFT and RIGHT. Many children find the fact that these commands can be assigned any number an exhilarating source of power and an exciting area of exploration. Deborah found it frightening, the reaction she had to most of what she did at school. In her first few hours of Turtle work she developed a disturbing degree of dependence on the instructor, constantly asking for reassurance before taking the smallest exploratory step. A turning point came when Deborah decided to restrict her Turtle commands, creating a microworld within the microworld of Turtle commands. She allowed herself only one turning command: RIGHT 30. To turn the Turtle through 90 degrees, she would repeat RIGHT 30 three times and would obtain the effect of LEFT 30 by repeating it eleven times. To an onlooker it might seem tedious to obtain simple effects in such complicated ways. But for Deborah it was exciting to be able to construct her own microworld and to discover how much she could do within its rigid constraints. She no longer asked permission to explore. And one day, when the teacher offered to show her a "simpler way" to achieve an effect, she listened patiently and said, "I don't think I'll do it that way." She emerged when she was ready, several weeks later, with a new sense of confidence that showed itself not only in more ambitious Turtle projects but in her relationship to everything else she did in school. I like to see in Deborah's experience a small recapitulation of how the success of such thinkers as Copernicus and Galileo allowed people to break away from superstitious dependencies that had nothing in themselves to do with physics. In both cases—in Deborah's personal history and in the history of Western thought—the success of a mathematical theory served more than an instrumental role: It served as an affirmation of the power of ideas and the power of the mind.

Chapter 5

Microworlds: Incubators for Knowledge

I HAVE DEFINED mathetics as being to learning as heuristics is to problem solving: Principles of mathetics are ideas that illuminate and facilitate the process of learning. In this chapter we focus on two important mathetic principles that are part of most people's common-sense knowledge about what to do when confronted with a new gadget, a new dance step, a new idea, or a new word. First, relate what is new and to be learned to something you already know. Second, take what is new and make it your own: Make something new with it, play with it, build with it. So for example, to learn a new word, we first look for a familiar "root" and then practice by using the word in a sentence of our own construction.

We find this two-step dictum about how to learn in popular, common-sense theories of learning: The procedure described for learning a new word has been given to generations of elementary schoolchildren by generations of parents and teachers. And it also corresponds to the strategies used in the earliest processes of learning. Piaget has studied the spontaneous learning of children and found both steps at work—the child absorbs the new into the old in a process that Piaget calls *assimilation,* and the child constructs his knowledge in the course of actively working with it.

But there are often roadblocks in the process. New knowledge often contradicts the old, and effective learning requires strategies to deal with such conflict. Sometimes the conflicting pieces of knowledge can be reconciled, sometimes one or the other must be abandoned, and sometimes the two can both be "kept around" if safely maintained in separate mental compartments. We shall look at these learning strategies by examining a particular case in which a formal theory of physics enters into sharp conflict with common-sense, intuitive ideas about physics.

One of the simplest of such conflicts is raised by the fundamental tenet of Newton's physics: A body in motion will, if left alone, continue to move forever at a constant speed and in a straight line. This principle of "perpetual motion" contradicts common experience and, indeed, older theories of physics such as Aristotle's.

Suppose we want to move a table. We apply a force, set the table in motion, and keep on applying the force until the table reaches the desired position. When we stop pushing, the table stops. To our superficial gaze, the table does not behave like a Newtonian object. If it did, textbooks tell us, one push would set it in motion forever and a counteracting force would be needed to stop it at the desired place.

This conflict of ideal theory and everyday observation is only one of several roadblocks to the learning of Newtonian physics. Others derive from difficulties in applying the two mathetic principles. According to the first, people who want to learn Newtonian physics should find ways to relate it to something they already know. But they may not possess any knowledge to which it can be effectively related. According to the second, a good strategy for learning would be to work with the Newtonian laws of motion, to use them in a personal and playful fashion. But this too is not so simple. One cannot do anything with Newton's laws unless one has some way to grab hold of them and some familiar material to which they can be applied.

The theme of this chapter is how computational ideas can serve as material for thinking about Newton's laws. The key idea has already been anticipated. We saw how formal geometry becomes more accessible when the Turtle instead of the point is taken as the

121

building block. Here we do for Newton what we did for Euclid. Newton's laws are stated using the concept of "a particle," a mathematically abstract entity that is similar to a point in having no size but that does have some other properties besides position: It has *mass* and *velocity* or, if one prefers to merge these two, it has *momentum*. In this chapter we enlarge our concept of Turtle to include entities that behave like Newton's particles as well as those we have already met that resemble Euclid's points. These new Turtles, which we call Dynaturtles, are more dynamic in the sense that their state is taken to include two velocity components in addition to the two geometric components, position and heading, of the previously discussed geometry Turtles. And having more parts to the state leads to requiring a slightly richer command language: TURTLE TALK is extended to allow us to tell the Turtle to set itself moving with a given velocity. This richer TURTLE TALK immediately opens up many perspectives besides the understanding of physics. Dynaturtles can be put into patterns of motion for aesthetic, fanciful, or playful purposes in addition to simulating real or invented physical laws. The too narrowly focused physics teacher might see all this as a waste of time: The real job is to understand physics. But I wish to argue for a different philosophy of physics education. It is my belief that learning physics consists of bringing physics knowledge into contact with very diverse personal knowledge. And to do this we should allow the learner to construct and work with transitional systems that the physicist may refuse to recognize as physics.[1]

Most physics curricula are similar to the math curriculum in that they force the learner into dissociated learning patterns and defer the "interesting" material past the point where most students can remain motivated enough to learn it. The powerful ideas and the intellectual aesthetic of physics is lost in the perpetual learning of "prerequisites." The learning of Newtonian physics can be taken as an example of how mathetic strategies can become blocked and unblocked. We shall describe a new "learning path" to Newton that gets around the block: a computer-based interactive learning environment where the prerequisites are built into the system and where learners can become the active, constructing architects of their own learning.

Let us begin with a closer look at the problem of prerequisites. Someone who wanted to learn about aerodynamics might lose interest upon seeing the set of prerequisites including mechanics and hydrodynamics that follow an exciting course description in a college catalogue. If one wants to learn about Shakespeare, one finds no list of prerequisites. It seems fair to assume that a list of prerequisites is an expression of what educators believe to be a learning path into a domain of knowledge. The learning path into aerodynamics is mathematical, and, as we have seen in our culture, mathematical knowledge is bracketed, treated as "special"—spoken of only in special places reserved for such esoteric knowledge. The nonacademic learning environments of most children provide little impetus to that mathematical development. This means that schools and colleges must approach the knowledge of aerodynamics along exceedingly formal learning paths. The route into Shakespeare is no less complex, but its essential constitutive elements are part of our general culture: It is assumed that many people will be able to learn them informally. The physics microworld we shall develop, the physics analog of our computer-based Mathland, offers a Piagetian learning path into Newtonian laws of motion, a topic usually considered paradigmatic of the kind of knowledge that can only be reached by a long, formalized learning path. Newtonian thinking about motion is a complex and seemingly counterintuitive set of assumptions about the world. Historically, it was long to evolve. And in terms of individual development, the child's interaction with his environment leads him to a very different set of personal beliefs about motion, beliefs that in many ways are closer to Aristotle's than to Newton's. After all, the Aristotelian idea of motion corresponds to the most common situation in our experience. Students trying to develop Newtonian thinking about motion encounter three kinds of problems that a computer microworld could help solve. First, students have had almost no direct experience of pure Newtonian motion. Of course, they have had some. For example, when a car skids on an icy road it becomes a Newtonian object: It will, only too well, continue in its state of motion without outside help. But the driver is not in a state of mind to benefit from the learning experience. In the absence of *direct* and *physical* experiences of Newtonian motion, the schools are forced to give the student indirect and highly mathematical experiences of Newtonian objects. There movement is learned by manipulating equations rather than by manipulating the objects themselves. The experience, lacking immediacy, is slow to change the student's intuitions. And it itself requires other formal prerequisites. The student must first learn how to work with equations before using them to model a Newtonian world. The simplest way in which our computer microworld might help is by putting students in a simulated world where they have direct access to Newtonian motion. This can be done when they are young. It need not wait for their mastery of equations. Quite the contrary: Instead of making students wait for equations, it can motivate and facilitate their acquisition of equational skills by providing an intuitively well understood context for their use.

Direct experience with Newtonian motion is a valuable asset for the learning of Newtonian physics. But more is needed to understand it than an intuitive, seat-of-the-pants experience. The student needs the means to conceptualize and "capture" this world. Indeed, a central part of Newton's great contribution was the invention of a formalism, a mathematics suited to this end. He called it "fluxions"; present-day students call it "differential calculus." The Dynaturtle on the computer screen allows the beginner to play with Newtonian objects. The concept of the Dynaturtle allows the student to think about them. And programs governing the behavior of Dynaturtles provide a formalism in which we can capture our otherwise too fleeting thoughts. In doing so it bypasses the long route (arithmetic, algebra, trigonometry, calculus) into the formalism that has passed with only superficial modification from Newton's own writing to the modern textbook. And I believe it brings the student in closer touch with what Newton must have thought before he began writing equations.

The third prerequisite is somewhat more subtle. We shall soon look directly at statements of what is usually known as Newton's laws of motion. As we do, many readers will no doubt recall a sense of unease evoked by the phrase "law of motion." What kind of a thing is that? What other laws of motion are there besides Newton's? Few students can answer these questions when they first encounter Newton, and I believe that this goes far toward explaining the difficulty of physics for most learners. Students cannot make a thing their own without knowing what kind of a thing it is. Therefore, the third prerequisite is that we must find ways to facilitate the personal appropriation not only of Newtonian motion and the laws that describe it, but also of the general notion of laws that describe motion. We do this by designing a series of microworlds.

The Turtle World was a microworld, a "place," a "province of Mathland," where certain kinds of mathematical thinking could hatch and grow with particular ease. The microworld was an incubator. Now we shall design a microworld to serve as an incubator for Newtonian physics. The design of the microworld makes it a "growing place" for a specific species of powerful ideas or intellectual structures. So, we design microworlds that exemplify not only the "correct" Newtonian ideas, but many others as well: the historically and psychologically important Aristotelian ones, the more complex Einsteinian ones, and even a "generalized law-of-motion world" that acts as a framework for an infinite variety of laws of motion that individuals can invent for themselves. Thus learners can progress from Aristotle to Newton and even to Einstein via as many intermediate worlds as they wish. In the descriptions that follow, the mathetic obstacles to Newton are overcome: The prerequisites are rooted in personal knowledge and the learner is involved in a creative exploration of the idea and the variety of laws of motion.

Let us begin to describe the microworld by starting with Newton's three laws, stated here "formally" and in a form that readers do not have to understand in detail:

1. Every particle continues in a state of rest or motion with constant speed in a straight line unless compelled by a force to change that state.
2. The net unbalanced force (F) producing a change of motion is equal to the product of the mass (m) and the acceleration (a) of the particle: $F=ma$.
3. All forces arise from the interaction of particles, and whenever a particle acts on another there is an equal and opposite reaction on the first.

As we have noted, children's access to these laws is blocked by more than the recondite language used to state them. We analyze these roadblocks in order to infer design criteria for our microworld. A first block is that children do not know anything else like these laws. Before being receptive to Newton's laws of motion, they should know some other laws of motion. There must be a first example of laws of motion, but it certainly does not have to be as complex, subtle, and counterintuitive as Newton's laws. More sensible is to let the learner acquire the concept of laws of motion by working with a very simple and accessible instance of a law of motion. This will be the first design criterion for our microworld. The second block is that the laws, as stated, offer no footholds for learners who want to manipulate them. There is no use they can put them to outside of end-of-chapter schoolbook exercises. And so, a second design criterion for our microworlds is the possibility of activities, games, art, and so on, that make activity in the microworlds matter. A third block is the fact that the Newtonian laws use a number of concepts that are outside most people's experience, the concept of "state," for example. Our microworld will be designed so that all needed concepts can be defined within the experience of that world.

As in the case of the geometry Turtle, the physics Turtle is an interactive being that can be manipulated by the learner, providing an environment for active learning. But the learning is not "active" simply in the sense of interactive. Learners in a physics microworld are able to invent their own personal sets of assumptions about the microworld and its laws and are able to make them come true. They can shape the reality in which they will work for the day, they can modify it and build alternatives. This is an effective way to learn, paralleling the way in which each of us once did some of our most effective learning. Piaget has demonstrated that children learn fundamental mathematical ideas by first building their own, very much different (for example, preconservationist) mathematics. And children learn language by first learning their own ("baby-talk") dialects. So, when we think of microworlds as incubators for powerful ideas, we are trying to draw upon this effective strategy: We allow learners to learn the "official" physics by allowing them the freedom to invent many that will work in as many invented worlds.

Following Polya's principle of understanding the new by associating it with the old, let us reinterpret our microworld of Turtle geometry as a microworld of a special kind of physics. We recast the laws by which Turtles work in a form that parallels the Newtonian laws. This gives us the following "Turtle laws of motion." Of course, in a world with only one Turtle, the third law, which deals with the interaction among particles, will not have an analog.

1. Every Turtle remains in its state of rest until compelled by a TURTLE COMMAND to change that state.
2. a. The input to the command FORWARD is equal to the Turtle's change in the POSITION part of its state.
b. The input to the command RIGHT TURN is equal to the Turtle's change of the HEADING part of its state.

What have we gained in our understanding of Newtonian physics by this exercise? How can students who know Turtle geometry (and can thus recognize its restatement in Turtle laws of motion) now look at Newton's laws? They are in a position to formulate in a qualitative and intuitive form the substance of Newton's first two laws by comparing them with something they already know. They know about states and state-change operators. In the Turtle world, there is a state-change operator for each of the two components of the state. The operator FORWARD changes the position. The operator TURN changes the heading. In physics, there is only one state-change operator, called *force*. The effect of force is to change velocity (or, more precisely, momentum). Position changes by itself.

These contrasts lead students to a qualitative understanding of Newton. Although there remains a gap between the Turtle laws and the Newtonian laws of motion, children can appreciate the second through an understanding of the first. Such children are already a big step ahead in learning physics. But we can do more to close the gap between Turtle and Newtonian worlds. We can design other Turtle microworlds in which the laws of motion move toward a closer approximation of the Newtonian situation.

127

MINDSTORMS

To do this we create a class of Turtle microworlds that differs in the properties that constitute the state of the Turtle and in the operators that change these states. We have formally described the geometry Turtle by saying that its state consists of position and velocity and that its state-change operators act independently of these two components. But there is another way, perhaps a more powerful and intuitive way, to think about it. This is to see the Turtle as a being that "understands" certain kinds of communication and not others. So, the geometry Turtle understood the command to change its position while keeping its heading and to change its heading while keeping its position. In the same spirit, we could define a Newtonian Turtle as a being that can accept only one kind of order, one that will change its momentum. These kinds of description are in fact the ones we use in introducing children to microworlds. Now let us turn to two Turtle microworlds that can be said to lie between the geometry and Newtonian Turtles.

VELOCITY TURTLES

The state of a velocity Turtle is POSITION AND VELOCITY. Of course, since velocity is defined as a change in position, by definition the first component of this state is continuously changing (unless VELOCITY is zero). So, in order to control a velocity Turtle, we only have to tell it what velocity to adopt. We do this by one state-change operator, a command called SETVELOCITY.

ACCELERATION TURTLES

Another Turtle, intermediate between the geometry Turtle and the Turtle that could represent a Newtonian particle, is an acceleration Turtle. Here, too, the state of the Turtle is its position and velocity. But this time the Turtle cannot understand such a command as "Take on such-and-such a velocity". It can only take instructions of the form "Change your velocity by x, no matter what your velocity happens to be." This Turtle behaves like a Newtonian particle with an unchangeable mass.

Thus, the sequence of turtles—geometry Turtle to velocity Turtle to acceleration Turtle to Newtonian Turtle—constitutes a path into Newton that is resonant with our two mathetic principles. Each step builds on the one before in a clear and transparent way, satisfying the principle of prerequisites. As for our second mathetic principle—"use it, play with it"—the case is even more dramatic. Piaget showed us how the child constructs a preconservationist and then a conservationist world out of the materials (tactile, visual, and kinesthetic) in his environment. But until the advent of the computer, there were only very poor environmental materials for the construction of a Newtonian world. However, each of the microworlds we described can function as an explorable and manipulable environment.

In Turtle geometry, geometry was taught by way of computer graphics projects that produce effects like those shown in the designs illustrating this book. Each new idea in Turtle geometry opened new possibilities for action and could therefore be experienced as a source of personal power. With new commands such as SETVELOCITY and CHANGE VELOCITY, learners can set things in motion and produce designs of ever-changing shapes and sizes. They now have even more personal power and a sense of "owning" dynamics. They can do computer animation—there is a new, personal relationship to what they see on television or in a pinball gallery. The dynamic visual effects of a TV show, an animated cartoon, or a video game now encourage them to ask how they could make what they see. This is a different kind of question than the one students traditionally answer in their "science laboratory." In the traditional laboratory pedagogy, the task posed to the children is to establish a given truth. At best, children learn that "this is the way the world works." In these dynamic Turtle microworlds, they come to a different kind of understanding—a feel for *why* the world works as it does. By trying many different laws of motion, children will find that the Newtonian ones are indeed the most economical and elegant for moving objects around.

All of the preceding discussion has dealt with Newton's first two laws. What analogs to Newton's third law are possible in the world of Turtles? The third law is only meaningful in a microworld of interactive entities—particles for Newton, Turtles for us. So let us assume a microworld with many Turtles that we shall call TURTLE 1, TURTLE 2, and so on. We can use TURTLE TALK to communicate with multiple Turtles if we give each of them a name. So we can use commands such as: TELL TURTLE 4 SETVELOCITY 20 (meaning "Tell Turtle number 4 to take on a velocity of 20).

Newton's third law expresses a model of the universe, a way to conceptualize the workings of physical reality as a self-perpetuating machine. In this vision of the universe, all actions are governed by particles exerting forces on one another, with no intervention by any outside agent. In order to model this in a Turtle microworld, we need many Turtles interacting with each other. Here we shall develop two models for thinking about interacting Turtles: linked Turtles and linked Dynaturtles.

In the first model we think of the Turtles as giving commands *to one another* rather than obeying commands from the outside. They are *linked Turtles*. Of course, Turtles can be linked in many ways. We can make Turtles that directly simulate Newtonian particles linked by simulated gravity. This is commonly done in LOGO laboratories, where topics usually considered difficult in college physics are translated into a form accessible to junior high school students. Such simulations can serve as a springboard from an elementary grasp of Newtonian mechanics to an understanding of the motion of planets and of the guidance of spacecraft. They do this by making working with the Newtonian principles an active and personally involving process. But to "own" the idea of interacting particles—or "linked Turtles"—the learner needs to do more. It is never enough to work within a given set of interactions. The learner needs to know more than one example of laws of interaction and should have experience inventing new ones. What are some other examples of linked Turtles?

A first is a microworld of linked Turtles called "mirror Turtles." We begin with a "mirror Turtle" microworld containing two Turtles linked by the rules: Whenever either is given a FORWARD (or BACK) command, the other does the same; whenever either is given a RIGHT TURN (or LEFT TURN) command, the other does the opposite. This means that if the two Turtles start off facing one another, any Turtle program will cause their trips to be mirror images of one another's. Once the learner understands this principle, attractive Kaleidoscope designs can easily be made.

A second microworld of linked Turtles, and one that is closer to Newtonian physics, applies these mirror linkages to velocity Turtles. No static images printed on this page could convey the visual excitement of these dynamic kaleidoscopes in which brightly colored points of light dance in changing and rotating paths. The end product has the excitement of art, but the process of making it involves learning to think in terms of the actions and reactions of linked moving objects.

These linked Turtle microworlds consolidate the learner's experience of the three laws of motion. But we have asserted that multiple microworlds also provide a platform for understanding the *idea* of a law of motion. A student who has mastered the general concept of a law of motion has a new, powerful tool for problem solving. Let's illustrate with the Monkey Problem.

> A monkey and a rock are attached to opposite ends of a rope that is hung over a pulley. The monkey and the rock are of equal weight and balance one another. The monkey begins to climb the rope. What happens to the rock?

I have presented this problem to several hundred MIT students, all of whom had successfully passed rigorous and comprehensive introductory physics courses. Over three quarters of those who had not seen the problem before gave incorrect answers or were unable to decide how to go about solving it. Some thought the position of the rock would not be affected by the monkey's climbing because the monkey's mass is the same whether he is climbing or not; some thought that the rock would descend either because of a conservation of energy or because of an analogy with levers; some guessed it would go up, but did not know why. The problem is clearly "hard." But this does not mean that it is "complex." I suggest that its difficulty is explicable by the lack of something quite simple. When they approach the problem, students ask themselves: "Is this a 'conservation-of-energy' problem?" "Is this a 'lever-arm' problem?" and so on. They do not ask themselves: "Is this a 'law-of-motion' problem?" They do not think in terms of such a category. In the mental worlds of most students, the concepts of conservation, energy, lever-arm, and so on, have become tools to think with. They are powerful ideas that organize thinking and problem solving. For a student who has had experience in a "laws-of-motion" microworld this is true of "law of motion." Thus this student will not be blocked from asking the right question about the monkey problem. It *is* a law-of-motion problem, but a student who sees laws of motion only in terms of algebraic formulas will not even ask the question. For those who pose the question, the answer comes easily. And once one thinks of the monkey and the rock as linked objects, similar to the ones we worked with in the Turtle microworld, it is obvious that they must both undergo the same changes in state. Since they start with the same velocity, namely zero, they must therefore always have the same velocity. Thus, if one goes up, the other goes up at the same speed.[2]

We have presented microworlds as a response to a pedagogical problem that arises from the structure of knowledge: the problem of prerequisites. But microworlds are a response to another sort of problem as well, one that is not embedded in knowledge but in the individual. The problem has to do with finding a context for the construction of "wrong" (or, rather, "transitional") theories. All of us learn by constructing, exploring, and theory building, but most of the theory building on which we cut our teeth resulted in theories we would have to give up later. As preconservationist children, we learned how to build and use theories only because we were allowed to hold "deviant" views about quantities for many years. Children do not follow a learning path that goes from one "true position" to another, more advanced "true position." Their natural learning paths include "false theories" that teach as much about theory building as true ones. But in school false theories are no longer tolerated.

Our educational system rejects the "false theories" of children, thereby rejecting the way children really learn. And it also rejects discoveries that point to the importance of the false-theory learning path. Piaget has shown that children hold false theories as a necessary part of the process of learning to think. The unorthodox theories of young children are not deficiencies or cognitive gaps, they serve as ways of flexing cognitive muscles, of developing and working through the necessary skills needed for more orthodox theorizing. Educators distort Piaget's message by seeing his contribution as revealing that children hold false beliefs, which they, the educators, must overcome. This makes Piaget-in-the-schools a Piaget backward—backward because children are being force-fed "correct" theories before they are ready to invent them. And backward because Piaget's work puts into question the idea that the "correct" theory is superior as a learning strategy.

Some readers may have difficulty seeing the child's nonconservationist view of the world as a kind of theory building. Let's take another example. Piaget asked preschool children, "What makes the wind?" Very few said, "I don't know." Most children gave their own personal theories, such as, "The trees made the wind by waving their branches." This theory, although wrong, gives good evidence for highly developed skill in theory building. It can be tested against empirical fact. Indeed there is a strong correlation between the presence of wind and the waving of tree branches. And children can perform an experiment that makes their causal connection quite plausible. When they wave their hands near their faces, they make a very noticeable breeze. Children can imagine this effect multiplied when the waving object is not a small hand but a giant tree, and when not one but many giant trees are waving. So, the trees of a dense forest should be a truly powerful wind generator.

What do we say to a child who has made such a beautiful theory? "That's great thinking, Johnny, but the theory is wrong" constitutes a put-down that will convince most children that making one's own theories is futile. So, rather than stifling the children's creativity, the solution is to create an intellectual environment less dominated than the school's by the criteria of true and false.

We have seen that microworlds are such environments. Just as students who prefer to do their programming using Newtonian Turtles with third law interaction are making Newton their own, children making a spectacular spiral in a *non-Newtonian* microworld are no less firmly on the path toward understanding Newton.

Both are learning what it is like to work with variables, to think in terms of ratios of dissimilar qualities, to make appropriate approximations, and so on. They are learning mathematics and science in an environment where true or false and right or wrong are not the decisive criteria.

As in a good art class, the child is learning technical knowledge as a *means* to get to a creative and personally defined end. There will be a product. And the teacher as well as the child can be genuinely excited by it. In the arithmetic class the pleasure that the teacher shows at the child's achievement is genuine, but it is hard to imagine teacher and child sharing delight over a product. In the LOGO environment it happens often. The spiral made in the Turtle microworld is a new and exciting creation by the child—he may even have "invented" the way of linking Turtles on which it is based.

The teacher's genuine excitement about the product is communicated to children who know they are doing something consequential. And unlike in the arithmetic class, where they know that the sums they are doing are just exercises, here they can take their work seriously. If they have just produced a circle by commanding the Turtle to take a long series of short forward steps and small right turns, they are prepared to argue with a teacher that a circle is really a polygon. No one who has overheard such a discussion in fifth-grade LOGO classes walks away without being impressed by the idea that the truth or falsity of theory is secondary to what it contributes to learning.

Chapter 6

Powerful Ideas in Mind-Size Bites

"I love your microworlds but is it physics? I don't say it is not. But how can I decide?"
—*A teacher*

A COMMON DISTINCTION between two ways of knowing is often expressed as "knowing-that" versus "knowing-how" or as "propositional knowledge" versus "procedural knowledge" or again as "facts" versus "skills." In this chapter we talk about some of the many kinds of knowing that cannot be reduced to either term of this dichotomy. Important examples from everyday life are knowing a person, knowing a place, and knowing one's own states of mind. In pursuit of our theme of using the computer to understand scientific knowing as rooted in personal knowing, we shall next look at ways in which scientific knowledge is more similar to knowing a person than similar to knowing a fact or having a skill. In this, we shall be doing something similar to how we used the Turtle to build bridges between formal geometry and the body geometry of the child. Here, too, our goal is to design conditions for more syntonic kinds of learning than those favored by the traditional schools. In previous chapters we have explored a paradox: Although most of our society classifies mathematics as the least accessible kind of knowledge, it is, paradoxically, the most accessible to children. In

135

this chapter we shall encounter a similar paradox in the domain of science. We shall look at ways in which the thinking of children has more in common with "real science" than "school science" has with the thinking either of children or of scientists. And once more we shall note a double paradox in the way computers enter into and influence this state of affairs. The introduction of the computer can provide a way out of the paradoxes, but it usually is used in ways that exacerbate them by reinforcing the paradoxical ways of thinking about knowledge, of thinking about "school math" and "school science."

Mathetically sophisticated adults use certain metaphors to talk about important learning experiences. They talk about *getting to know* an idea, *exploring* an area of knowledge, and *acquiring sensitivity* to distinctions that seemed ungraspably subtle just a little while ago.

I believe that these descriptions apply very accurately to the way children learn. But when I asked students in grade schools to talk about learning, they used a very different kind of language, referring mainly to facts they had learned and skills they had acquired. It seems very clear that school gives students a particular model of learning; I believe it does this not only through its way of talking but also through its practices.

Skills and the discrete facts are easy to give out in controlled doses. They are also easier to measure. And it is certainly easier to enforce the learning of a skill than it is to check whether someone has "gotten to know" an idea. It is not surprising that schools emphasize learning skills and facts and that students pick up an image of learning as "learning that" and "learning how."

Working in Turtle microworlds is a model for what it is to get to know an idea the way you get to know a person. Students who work in these environments certainly do discover facts, make propositional generalizations, and learn skills. But the primary learning experience is not one of memorizing facts or of practicing skills. Rather, it is getting to know the Turtle, exploring what a Turtle can and cannot do. It is similar to the child's everyday activities, such as making mudpies and testing the limits of parental authority—all of which have a component of "getting to know." Teachers often set up situations in which they claim that children are actually getting to know this or that concept even though they might not realize it. Yet the Turtle is different—it allows children to be deliberate and conscious in bringing a kind of learning with which they are comfortable and familiar to bear on math and physics. And, as we have remarked, this is a kind of learning that brings the child closer to the mathetic practice of sophisticated adult learners. The Turtle in all its forms (floor Turtles, screen Turtles and Dynaturtles) is able to play this role so well because it is both an engaging anthropomorphizable object and a powerful mathematical idea. As a model for what mathematical and scientific learning is about, it stands in sharp contrast to the methodology described by the fifth grader, Bill (mentioned in chapter 3), who told me that he learned math by making his mind a blank and saying it over and over.

For me, getting to know a domain of knowledge (say, Newtonian mechanics or Hegelian philosophy) is much like coming into a new community of people. Sometimes one is initially overwhelmed by a bewildering array of undifferentiated faces. Only gradually do the individuals begin to stand out. On other occasions one is fortunate in quickly getting to know a person or two with whom an important relationship can develop. Such good luck may come from an intuitive sense for picking out the "interesting" people, or it may come from having good introductions. Similarly, when one enters a new domain of knowledge, one initially encounters a crowd of new ideas. Good learners are able to pick out those who are powerful and congenial. Others who are less skillful need help from teachers and friends. But we must not forget that while good teachers play the role of mutual friends who can provide introductions, the actual job of getting to know an idea or a person cannot be done by a third party. Everyone must acquire skill at getting to know and a personal style for doing it.

Here we use an example from physics to focus the image of a domain of knowledge as a community of powerful ideas, and in doing so take a step toward an epistemology of powerful ideas. Turtle microworlds illustrate some general strategies for helping a newcomer begin to make friends in such a community. A first strategy is to ensure that the learner has a model for this kind of learning;

working with Turtles is a good one. This strategy does not require that all knowledge be "Turtle-ized" or "reduced" to computational terms. The idea is that early experience with Turtles is a good way to "get to know" what it is like to learn a formal subject by "getting to know" its powerful ideas. I made a similar point in chapter 2 when I suggested that Turtle geometry could be an excellent domain for introducing learners to Polya's ideas about heuristics. This does not make heuristic thinking dependent on turtles or computers. Once Polya's ideas are thoroughly "known," they can be applied to other domains (even arithmetic). Our discussion in chapter 4 suggested that theoretical physics may be a good carrier for an important kind of meta-knowledge. If so, this would have important consequences for our cultural view of its role in the lives of children. We might come to see it as a subject suitable for early acquistion not simply because it explicates the world of things but because it does so in a way that places children in better command of their own learning processes.

For some people taking physics as a model for how to analyze problems is synonymous with a highly quantitative, formalistic approach. And indeed, the story of what has happened when such disciplines as psychology and sociology have taken physics as a model has often had unhappy endings. But there is a big difference in the kind of physics used. The physics that had a bad influence on social sciences stressed a positivistic philosophy of science. I am talking about a kind of physics that places us in firm and sharp opposition to the positivistic view of science as a set of true assertions of fact and of "law." The propositional content of science is certainly very important, but it constitutes only a part of a physicist's body of knowledge. It is not the part that developed first historically, it is not a part that can be understood first in the learning process, and it is, of course, not the part I am proposing here as a model for reflection about our own thinking. We shall be interested in knowledge that is more qualitative, less completely specified, and seldom stated in propositional form. If students are given such equations as $f = ma$, $E = IR$, or $PV = RT$ as the primary models of the knowledge that constitutes physics, they are placed in a position where nothing in their own heads is likely to be recognized as "physics."

We have already seen that this is the kind of thing that puts them at very high risk as learners. They are on the road to dissociated learning. They are on the road to classifying themselves as incapable of understanding physics. A different sense of what kind of knowledge constitutes physics is obtained by working with Turtles: Here a child, even a child who possesses only one piece of fragmentary, incompletely specified, qualitative knowledge (such as "these Turtles only understand changing velocities") can already do something with it. In fact, he or she can start to work through many of the conceptual problems that plague college students. The fragment of knowledge can be used without even knowing how to represent velocities quantitatively! It is of a kind with the intuitive and informal but often very powerful ideas that inhabit all of our heads whether we are children or physicists.

This use of the computer to create opportunities for the exercise of qualitative thinking is very different from the use of computers that has become standard in high school physics courses. There it is used to reinforce the quantitative side of physics by allowing more complex calculations. Thus it shares some of the paradox we have already noted in the use of new technologies to reinforce educational methods whose very existence is a reflection of the limitations of the precomputer period. As previously mentioned, the need for drill and practice in arithmetic is a symptom of the absence of conditions for the syntonic learning of mathematics. The proper use of computers is to supply such conditions. When computers are used to cure the immediate symptom of poor scores in arithmetic, they reinforce habits of dissociated learning. And these habits which extend into many areas of life are a much more serious problem than weakness in arithmetic. The cure may be worse than the disease. There is an analogous argument about physics. Traditional physics teaching is forced to overemphasize the quantitative by the accidents of a paper-and-pencil technology which favors work that can produce a definite "answer." This is reinforced by a teaching system of using "laboratories" where experiments are done to prove, disprove, and "discover" already known propositions. This makes it very difficult for the student to find a way to constructively bring together intuitions and formal methods. Everyone is too busy following the cookbook. Again, as in the case of arithmetic, the computer should be used to remove the fundamental problem. However, as things are today, the established image of school physics as quantitative and the established image of the computer reinforce each other. The computer is used to aggravate the already too-quantitative methodology of the physics classes. As in the case of arithmetic drill and practice, this use of the computer undoubtedly produces local improvements and therefore gets the stamp of approval of the educational testing community and of teachers who have not had the opportunity to see something better. But throughout this book we have been developing the elements of a less quantitative approach to computers in education. Now we directly address the concerns this shift in direction must raise for a serious teacher of physics.

The quotation at the beginning of this chapter was spoken in some anguish by a teacher who manifestly liked working with Turtles but could not reconcile it with what she had come to define as "doing physics". The situation reflects a permanent dilemma faced by anyone who wishes to produce radical innovation in education. Innovation needs new ideas. I have argued that we should be prepared to undertake far-reaching reconceptualizations of classical domains of knowledge. But how far can this go? Education has a responsibility to tradition. For example, the job of the community of English teachers must be to guide their students to the language and literature as it exists and as it developed historically. They would be failing in their duty if instead they invented a new language, wrote their version of poetry, and passed on to the next generation these fabricated entities in the place of the traditional ones. The concern of the teacher worried about whether working with Turtles is "really learning physics" is very serious.

Is work with Turtles analogous to replacing Shakespeare by "easier," made-up literature? Does it bring students into contact with the intellectual products of Galileo, Newton, and Einstein or merely with an idiosyncratic invention that is neither marked by greatness nor tested by time? The question raises fundamental problems, among them: What *is* physics? And what is the potential influence of computation on understanding it?

Most curriculum designers have easy answers to these questions. They define elementary physics as what is taught in schools. Occasionally they move material usually taught in college down to high school, or bring in new topics of the same kind as the old. For example, modern particles are mentioned and the textbooks show schematically how a nuclear reactor works. Even the more visionary curriculum reformers stayed within the conceptual framework defined by equations, quantitative laws, and laboratory experiments. Thus, they could feel secure that they were really "teaching physics." The possibility opened by the computer of a new kind of activity and of a new relationship to ideas poses problems of responsibility toward the cultural heritage. I take this responsibility seriously but cannot feel that I serve it by taking shelter behind the existing curriculum. One cannot accept this shelter without seriously considering the question of whether school science is not already in the position of the hypothetical English teacher who taught an ersatz form of English because it seemed to be more teachable. I believe that this is the case.

In chapter 5, I suggested that it is "school physics" rather than "Turtle physics" that betrays the spirit of "real physics." Here I pursue my argument by talking about components of physics that are even further removed than Dynaturtles from the traditional curriculum. These are very general, usually qualitative, intuitive ideas or "frames" used by physicists to think about problems before they can even decide what quantitative principles apply.

I ask readers who may not be familiar with such qualitative thinking in physics to follow a hypothetical conversation between two great physicists.

Many millions of students have grown up believing that Galileo refuted Aristotle's expectation that the time taken for an object to fall to the ground is proportional to its weight by dropping cannonballs from the tower of Pisa. Galileo's experiment is supposed to have proved that except for minor perturbations due to air resistance, a heavy and a light cannonball would, if dropped together, reach the ground together. In fact it is extremely unlikely that Galileo performed any such experiment. But whether he did or did not is less interesting than the fact that he would not have had the slightest doubt about the outcome of the experiment. In order to convey a sense of the kind of thinking that could have given him this assurance, we shall go through a hypothetical dialog between two imaginary characters, GAL and ARI.

> GAL: Look, your theory has got to be wrong. Here's a two-pound and a one-pound ball. The two-pound ball takes two seconds to fall to the ground. Tell me, how long do you think the one-pound ball would need?
>
> ARI: I suppose it would take four seconds. Anyway, much more than two seconds.
>
> GAL: I thought you would say that. But now please answer another question. I am about to drop two one-pounders simultaneously. How long will the pair of them take to reach the ground?
>
> ARI: That's not another question. I gave my opinion that one-pound balls take four seconds. Two of them must do the same. Each falls independently.
>
> GAL: You are consistent with yourself if two bodies are two bodies, not one.
>
> ARI: As they are . . . of course.
>
> GAL: But now if I connect them by a gossamer thread . . . is this now two bodies or one? Will it (or they) take two seconds or four to fall to the ground?
>
> ARI: I am truly confused. Let me think. . . . It's one body, but then it should fall for two seconds before reaching the earth. But then this would mean that a thread finer than silk could speed up a furiously falling ball of iron. It seems impossible. But if I say it is two bodies . . . I am in deep trouble. What is a body? How do I know when one becomes two? And if I cannot know then how sure can I be of my laws of falling bodies?

From a strictly logical point of view, GAL's argument is not absolutely compelling. One can imagine "fixes" for ARI's theory. For example, he could propose that the time taken might depend on the form as well as the weight of the body. This would allow him the possibility that a two-pound body made of two cannonballs and gossamer threads fall more slowly than a two-pound sphere of iron. But in fact the kind of argument used by GAL is subversive of the kind of theory expounded by ARI, and historically, it is highly plausible that the great conversion from Aristotelian thinking was fueled by such arguments. No single argument could by itself convert Aristotle, for whom the theory of falling objects was an element in a mutually supporting web. But as GAL's way of thinking gained currency, the Aristotelian system was eroded. Indeed I contend that arguments of this kind, as opposed to the apparently more compelling arguments from precise facts and equations, play an essential role in the evolution of thinking, both on the historical scale of the evolution of science itself and on the personal scale of the development of the individual learner.

ARI would have been far better able to defend himself had GAL argued from specific facts or calculations, which might allow quibbles about their conditions of applicability and allow themselves to be compartmentalized. The hard punch of GAL's argument comes from the fact that it mobilizes ARI's own intuitions about the nature of physical objects and about the continuity of natural effects (thinner than silk versus furiously falling iron). To a logician this argument might seem less compelling. But as empathetic fellow humans we find ourselves squirming in confusion with ARI.

There is a lot to be learned by thinking through the issues raised by this dialogue, simplistic as it is. First we note that GAL is not just being cleverer than ARI: He *knows* something that ARI seems not to know. In fact, if we look carefully we see that GAL skillfully deploys several powerful ideas. Most striking is his principal idea of looking at a two-pound object as made up of two one-pound objects, seeing the whole as additively made of whatever parts we care to divide it into. Stated abstractly this idea sounds trivial in some contexts and simply false in others: We are used to being reminded that "the whole is more than the sum of its parts." But we should not treat it as a proposition to be judged by the criterion of truth and falsity. It is an *idea,* an intellectual tool, and one that has proved itself to be enormously powerful when skillfully used.

GAL's idea is powerful and is part of the intellectual tool kit of every modern mathematician, physicist, or engineer. It is as important in the history and in the learning of physics as the kind of knowledge that fits into propositions or equations. But one would not know this from looking at textbooks. GAL's idea is not given a name, it is not attributed to a historical scientist, it is passed over in silence by teachers. Indeed, like most of intuitive physics, this knowledge seems to be acquired by *adult* physicists through a process of Piagetian learning, without, and often in spite of, deliberate classroom teaching. Of course, my interest in recognizing the existence of these informally learned, powerful intuitive ideas is not to remove them from the scope of Piagetian learning and place them in a curriculum: There are other ways to facilitate their acquisition. By recognizing their existence we should be able to create conditions that will foster their development, and we certainly can do a lot to remove obstacles that block them in many traditional learning environments.

GAL's dialogue with ARI has something to teach us about one of the most destructive blocks to learning: the use of formal reasoning to put down intuitions.

Everyone knows the unpleasant feeling evoked by running into a counterintuitive phenomenon where we are forced, by observation or by reason, to acknowledge that reality does not fit our expectations. Many people have this feeling when faced with the perpetual motion of a Newtonian particle, with the way a rudder turns a boat, or with the strange behavior of a toy gyroscope. In all these cases intuition seems to betray us. Sometimes there is a simple "fix"; we see that we made a superficial mistake. But the interesting cases are those where the conflict remains obstinately in place however much we ponder the problem. These are the cases where we are tempted to conclude that "intuition cannot be trusted." In these situations we need to improve our intuition, to debug it, but the pressure on us is to abandon intuition and rely on equations instead. Usually when a student in this plight goes to the physics teacher saying, "I think the gyroscope should fall instead of standing upright," the teacher responds by writing an equation to *prove* that the thing stands upright. But that is not what the student needed. He already *knew* that it would stay upright, and this knowledge hurt by conflicting with intuition. By *proving* that it will stand upright the teacher rubs salt in the wound but does nothing to heal it. What the student needs is something quite different: better understanding of himself, not of the gyroscope. He wants to know why his intuition gave him a wrong expectation. He needs to know how to work on his intuitions in order to change them. We see from the dialogue that GAL is an expert at how to manipulate intuitions. He does not force ARI into rejecting intuition in favor of calculation. Rather he forces him to confront a very specific aspect of his intuitive thinking: *how he thinks about objects.* One suspects from the dialogue that GAL is used to understanding objects by thinking of them as composed of parts, or subobjects, while ARI is used to thinking of objects more globally, as undivided wholes with global properties such as shape and weight.

We might seem to have strayed far from our discussion of computers. But the interaction between GAL and ARI is close to an important kind of interaction between children and computers and between children and instructors via computers. GAL tried to make ARI confront and work through his intuitive ways of thinking about objects, and ARI might be skillful enough to do so. But what can children do to confront *their* intuitions?

Of course the question is rhetorical in that I know that children think a great deal about their thinking. They do worry about their intuitions. They do confront them and they do debug them. If they did not the idea of making them do so would indeed be utopian. But since they do it already, we can provide materials to help them do it better.

I see the computer as helping in two ways. First, the computer allows, or obliges, the child to externalize intuitive expectations. When the intuition is translated into a program it becomes more obtrusive and more accessible to reflection. Second, computational ideas can be taken up as materials for the work of remodeling intuitive knowledge. The following analysis of a well-known puzzle is used to illustrate how a Turtle model can help bridge the gap between formal knowledge and intuitive understanding. We have seen many examples in incidents where children work with computers. Here I shall convey a sense of what this means by inviting you to work on a situation where *your* intuitions will come into conflict.

The purpose in working on the problem is not to "get the right answer," but to look sensitively for conflict between different ways of thinking about the problem: for example, between two intuitive ways of thinking or between an intuitive and a formal analysis. When you recognize conflicts, the next step is to work through them until you feel more comfortable. When I did this, I found that the Turtle model was extremely helpful in resolving some of the conflicts. But my reaction is undoubtedly shaped by my positive feelings about Turtles.

Imagine a string around the circumference of the earth, which for this purpose we shall consider to be a perfectly smooth sphere, four thousand miles in radius. Someone makes a proposal to place the string on six-foot-high poles. Obviously this implies that the string will have to be longer. A discussion arises about how much longer it would have to be. Most people who have been through high school know how to *calculate* the answer. But before doing so or reading on try to guess: Is it about one thousand miles longer, about a hundred, or about ten?

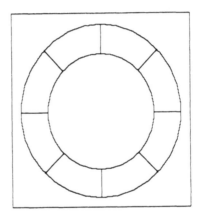

Figure 15

The figure shows a string around the earth supported by poles of greatly exaggerated height. Call the radius of the earth *R* and the height of the poles *h*. The problem is to estimate the difference in length between the outer circumference and the true circumference. This is easy to calculate from the formula:

CIRCUMFERENCE = $2\pi \times$ RADIUS

So the difference must be

$2\pi(R+h) - 2\pi R$

which is simply $2\pi h$.

But the challenge here is to "intuit" an approximate answer rather than to "calculate" an exact one.

Most people who have the discipline to think before calculating—a discipline that forms part of the know-how of debugging one's intuitions—experience a compelling intuitive sense that "a lot" of extra string is needed. For some the source of this conviction seems to lie in the idea that something is being added all around the twenty-four thousand miles (or so) of the earth's circumference. Others attach it to more abstract considerations of proportionality. But whatever the source of the conviction may be it is "incorrect" in anticipating the result of the formal calculation, which turns out to be a little less than forty feet. The conflict between intuition and calculation is so powerful that the problem has become widely known as a teaser. And the conclusion that is often drawn from this conflict is that intuitions are not to be trusted. Instead of drawing this conclusion, we shall attempt to engage the reader in a dialog in order to identify what needs to be done to alter this intuition.

As a first step we follow the principle of seeking out a similar problem that might be more tractable. And a good general rule for simplification is to look for a linear version. Thus we pose the same problem on the assumption of a "square earth."

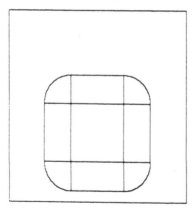

Figure 16a

The string on poles is assumed to be at distance *h* from the square. Along the edges the string is straight. As it goes around the corner it follows a circle of radius *h*. The straight segments of the string have the same length as the edges of the square. The extra length is all at the corners, in the four quarter-circle pie slices. The four quarter circles make a whole circle of radius *h*. So the "extra string" is the circumference of this circle, that is to say $2\pi h$.

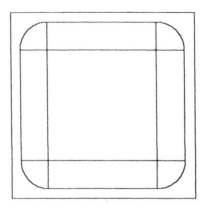

Figure 16b

Increasing the size of the square does not change the quarter-circle pie slices. So the extra string needed to raise a string from the ground to height *h* is the same for a very small square earth as for a very large one.

The diagram gives us a geometric way to see that the same amount of extra string is needed here as in the case of the circle. This is itself quite startling. But more startling is the fact that we can see so directly that the size of the square makes no difference to how much extra string is needed. We could have calculated this fact by formula. But doing so would have left us in the same difficulty. By "seeing" it geometrically we can bring this case into line with our intuitive principle: Extra string is needed only where the earth curves. Obviously no extra string is needed to raise a straight line from the ground to a six-foot height.

Unfortunately, this way of understanding the square case might seem to undermine our understanding of the circular case. We have completely understood the square but did so by seeing it as being very much different from the circle.

But there is another powerful idea that can come to the rescue. This is the idea of intermediate cases: *When there is a conflict between two cases, look for intermediates,* as GAL in fact did in constructing a series of intermediate objects between the two one-pound balls and one two-pound ball. But what is intermediate

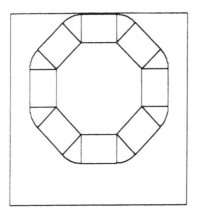

Figure 17
In the octagon, too, the "extra string" is all in the pie slices at the
corners. If you put them together they form a circle of radius *h*. As in
the case of the square, this circle is the same whether the octagon is
small or big. What works for the square (4-*gon*) and for the octagon
(8-*gon*) works for the 100-*gon* and for the 1000-*gon*.

between a square and a circle? Anyone who has studied calculus or
Turtle geometry will have an immediate answer: polygons with
more and more sides. So we look at Figure 17, which show strings
around a series of polygonal earths. We see that the extra string
needed remains the same in all these cases and, remarkably, we see
something that might erode the argument that the circle adds
something all around. The 1000-gon adds something at many more
places than the square, in fact two hundred fifty times as many
places. But it adds less, in fact one two hundred fiftieth at each of
them.

Now will your mind take the jump? Like GAL, I have said nothing so far to compel this crucial step by rigorous logic. Nor shall I.
But at this point some people begin to waver, and I conjecture that
whether they do or not depends on how firm a commitment they
have made to the idea of polygonal approximations to a circle. For
those who have made the polygonal representation their own, the
equivalence of polygon and circle is so immediate that intuition is
carried along with it. People who do not yet "own" the equivalence

between polygonal representation and circle can work at becoming better acquainted with it, for example, by using it to think through other problems.

The following problem is taken from Martin Gardner's book, *Mathematical Carnival:*

> 'If one penny rolls around another penny without slipping how many times will it rotate in making one revolution? One might guess the answer to be one, since the moving penny rolls along an edge equal to its own circumference, but a quick experiment shows that the answer is two; apparently the complete revolution of the moving penny adds an extra rotation.'[1]

Again there is a conflict between the intuitive guess (one revolution) and the result of more careful investigation. How can one bring one's intuition into line?

The same strategy works here as for the string around the earth problem. Roll a penny around a square without slipping. You will notice that it behaves quite differently as it rolls along the sides than when it pivots around the corners. It is easy to see that the total rotation at the four corners combined is 360°. This remains true for any polygon, however many sides it has and however big it is. And once more, the crucial step becomes the passage from the polygon to a Turtle circle to a true circle.

I am not suggesting that one more exercise will change your intuition of circularity. Here too, as in the case of Aristotle's physics, the particular piece of knowledge is part of a large network of mutually supportive ways of thinking. I am suggesting that you keep this new way of thinking in mind for awhile, looking for opportunities to use it as you might look for opportunities to introduce a new friend to old ones. And even then, I have no way of knowing whether you *want* to change your intuition of circularity. But if it is to change I think that the process I am suggesting here is the best, perhaps the only, way whether it is adopted deliberately or simply happens unconsciously.

I want you to go away from this book with a new sense of a child's value as a thinker, even as an "epistemologist" with a notion of the power of powerful ideas. But I also realize that these images might seem abstract and even irritating to some of you, perhaps especially those of you who teach children.

For example, a third-grade teacher who spends many frustrating hours every day trying to teach thirty-six children to write grammatical sentences and to do arithmetic might view my suggestions about Turtle geometry, physics microworlds, and cybernetics as far removed from reality, as far removed as was Marie Antoinette when she suggested that those who were starving for bread should eat cake. How are the powerful ideas we have discussed related to what most schools see as their bread-and-butter work, that is to say, the basic skills?

A first connection works through the attitude of the learner. You can't learn bread-and-butter skills if you come to them with fear and the anticipation of hating them. When children who will not let a number into their head fail to learn arithmetic, the remedy must be developing a new relationship with numbers. Achieving this can put children in a positive relationship to anything else that they will recognize as being of the same kind. This can be school mathematics.

> Kim was a fifth-grade girl who invariably came out on the bottom on all school arithmetic tests. She hated math. In a LOGO environment she became engrossed in programming. She designed a project that maintained a special database to store information about her family tree. One day a visiting educator remarked to her that "computers made math fun." Kim looked up from her work and said very angrily: "There ain't nothin' fun in math." The instructor in her class had not thought it advisable to discuss with her whether what she was doing with the computer was "math." Clearly, anything that was good was definitionally not math. But by the end of the year Kim made the connection herself and decided that mathematics was neither unpleasant nor difficult.

Getting to know (and like) mathematics as you get to know (and like) a person is a very pertinent image of what happened in this case. Computers can also contribute to the learning of bread-and-butter arithmetic by changing our perception of what it is about, of what powerful ideas are most important in it. School arithmetic, generally thought of as a branch of number theory, might better be thought of as a branch of computer science. Difficulties experienced by children are not usually due to deficiencies in their notion of number but in failing to appropriate the relevant algorithms. Learning algorithms can be seen as a process of making, using, and fixing programs. When one adds multidigit numbers one is in fact acting as a computer in carrying through a procedure something like the program in Figure 18.

1.	Set out numbers following conventional format.
2.	Focus attention on the rightmost column.
3.	Add as for single digit numbers.
4.	If result <10 record results.
5.	If result in rightmost column was equal to or greater than 10, then record rightmost digit and enter rest in next column to left.
6.	Focus attention one column to left.
7.	Go to line 3.

Figure 18

To get better at this sort of activity one needs to know more about, and feel more comfortable with, the ways of procedures. And this, of course, is what a good computer experience allows.

These remarks should be put in the context of our earlier discussion about the difference between the New Math curriculum reform of the 1960s and the kind of enrichment the computer culture can bring to mathematics. In chapter 2 we dealt with one important reason for the failure of the New Math: It did not ameliorate our society's alienated relationship with number. On the contrary, it aggravated it. We now see a second reason for the failure of the New Math. It tried to root the teaching of math in number theory, set theory, or logic instead of facing the conceptual stumbling blocks that children really experience: Their lack of knowledge about programming. Thus the authors of the New Math misunderstood the source of children's problems. This misunderstanding is harmful in several ways. It is harmful insofar as it seeks to improve the child's understanding of arithmetic by drill in irrelevant areas of knowledge. It is also harmful insofar as it imparts an inappropriate value system into mathematics education. The pure mathematician sees the idea of number as valuable, powerful, and important.

The details of procedure are seen as superficial and uninteresting. Thus the child's difficulties are referred back to abstract difficulties with the notion of number. The computer scientist takes a more direct approach. Trouble with adding is not seen as symptomatic of something else; it is trouble with the *procedure* of adding. For the computerist the procedure and the ways it can go wrong are fully as interesting and as conceptual as anything else. Moreover, *what* went wrong, namely the bugs, are not seen as mistakes to be avoided like the plague, but as an intrinsic part of the learning process.

Ken was a fifth grader who added 35 and 35 and got 610. His bug was showing clearly. Since 32 plus 32 is 64, then 35 plus 35 should be 610. Ken was brought into a better relationship with mathematics when he learned to see his mistake as a trick that mathematical formalisms play on us. The French *can* say seventy as *soixante dix,* "sixty-ten," but although they can write sixty-five as 65, they cannot write sixty-ten as 610. This symbol has been preempted to mean something else.

Ken might superficially appear to have had bad intuitions about numbers. But this is quite wrong as a diagnosis. When asked "If you had thirty-five dollars and you got thirty-five dollars more, would you have $610.00," his answer was an emphatic, "No way." When asked how much he would have, he returned to his paper calculation, crossed off the zero from 610, and came up with the new answer of 61, which intuitively is not so far off. His problem is not bad intuition or notion of number. From a computerist's point of view one can recognize several difficulties, each of which is understandable and correctable.

First, he dissociates the operation of the procedure from his general store of knowledge. A better procedure would have an "error check" built into it. Since he could recognize the error when prompted, he certainly should have been capable of setting up the procedure to include prompting himself. Second, when he found the error he did not change, or even look at, the procedure, but merely changed the answer. Third, my knowledge of Ken tells me why he did not try to change the procedure. At the time of this incident he did not recognize procedures as entities, as things one could name, manipulate, or change. Thus, fixing his procedures is very far indeed from his awareness. The idea of procedures as things that can be debugged is a powerful, difficult concept for many children, until they have accumulated experience in working with them.

I have seen children like Ken get over this kind of difficulty after some experience writing programs in a LOGO environment. But why don't children learn a procedural approach from daily life? Everyone works with procedures in everyday life. Playing a game or giving directions to a lost motorist are exercises in procedural thinking. But in everyday life procedures are lived and used, they are not necessarily reflected on. In the LOGO environment, a procedure becomes a thing that is named, manipulated, and recognized as the children come to acquire the idea of procedure. The effect of this for someone like Ken is that everyday-life experience of procedures and programming now becomes a resource for doing formal arithmetic in school. Newton's laws of motion came alive when we used computational metaphor to tie them to more personal and conceptually powerful things. Geometry came alive when we connected it to its precursors in the most fundamental human experience: the experience of one's body in space. Similarly, formal arithmetic will come alive when we can develop links for the individual learner with its procedural precursors. And these precursors do exist. The child does have procedural knowledge and he does use it in many aspects of his life, whether in planning strategies for a game of tic-tac-toe or in giving directions to a motorist who has lost his way. But all too often the same child does not use it in school arithmetic.

The situation is exactly like the one we met in the dialog between ARI and GAL and in the use of the Turtle circle model to change the intuition of circularity brought to bear on the string and coin problems. In all these cases, we are interested in how a powerful idea is made part of intuitive thinking. I do not know a recipe for developing a child's intuition about when and how to use procedural ideas, but I think that the best we can do is what is suggested by the metaphor of getting to know a new person. As educators we can help by creating the conditions for children to use procedural thinking effectively and joyfully. And we can help by giving them access to many concepts related to procedurality. This is achieved through the conceptual content of LOGO environments.

In this book I have clearly been arguing that procedural thinking is a powerful intellectual tool and even suggested analogizing oneself to a computer as a strategy for doing it. People often fear that using computer models for people will lead to mechanical or linear thinking: They worry about people losing respect for their intuitions, sense of values, powers of judgment. They worry about instrumental reason becoming a model for good thinking. I take these fears seriously but do not see them as fears about computers themselves but rather as fears about how culture will assimilate the computer presence. The advice "think like a computer" could be taken to mean *always* think about everything like a computer. This would be restrictive and narrowing. But the advice could be taken in a much different sense, not precluding anything, but making a powerful addition to a person's stock of mental tools. Nothing is given up in return. To suggest that one must give up an old method in order to adopt a new one implies a theory of human psychology that strikes me as naive and unsupported. In my view a salient feature of human intelligence is the ability to operate with many ways of knowing, often in parallel, so that something can be understood on many levels. In my experience, the fact that I ask myself to "think like a computer" does not close off other epistemologies. It simply opens new ways for approaching thinking. The cultural assimilation of the computer presence will give rise to a computer literacy. This phrase is often taken as meaning knowing how to program, or knowing about the varied uses made of computers. But true computer literacy is not just knowing how to make use of computers and computational ideas. It is knowing when it is appropriate to do so.

Chapter 7

Logo's Roots: Piaget and AI

THE READER has already met a variety of learning situations drawn together by a common set of ideas about what makes for effective learning. In this chapter we turn directly to these ideas and to the theoretical sources by which they are informed. Of these we focus on two: first, the Piagetian influence, and second, the influence of computational theory and artificial intelligence.

I have previously spoken of "Piagetian learning," the natural, spontaneous learning of people in interaction with their environment, and contrasted it with the curriculum-driven learning characteristic of traditional schools. But Piaget's contribution to my work has been much deeper, more theoretical and philosophical. In this chapter I will present a Piaget very different from the one most people have come to expect. There will be no talk of stages, no emphasis on what children at certain ages can or cannot learn to do. Rather I shall be concerned with Piaget the epistemologist, as his ideas have contributed toward the knowledge-based theory of learning that I have been describing, a theory that does not divorce the study of how mathematics is learned from the study of mathematics itself.

I think these epistemological aspects of Piaget's thought have been underplayed because up until now they offered no possibilities for action in the world of traditional education. But in a computer-rich educational environment, the educational environment of the next decade, this will not be the case. In chapter 5 and in the development of the Turtle idea itself we saw examples of how an epistemological inquiry into what is fundamental in a sector of mathematics, the mathematics of differential systems, has already paid off in concrete, effective educational designs. The Piaget of the stage theory is essentially conservative, almost reactionary, in emphasizing what children cannot do. I strive to uncover a more revolutionary Piaget, one whose epistemological ideas might expand known bounds of the human mind. For all these years they could not do so for lack of a means of implementation, a technology which the mathetic computer now begins to make available.

The Piaget as presented in this chapter is new in another sense as well. He is placed in a theoretical framework drawn from a side of the computer world of which we have not spoken directly, but whose perspectives have been implicit throughout this book, that of artificial intelligence, or AI. The definition of artificial intelligence can be narrow or broad. In the narrow sense, AI is concerned with extending the capacity of machines to perform functions that would be considered intelligent if performed by people. Its goal is to construct machines and, in doing so, it can be thought of as a branch of advanced engineering. But in order to construct such machines, it is usually necesary to reflect not only on the nature of machines but on the nature of the intelligent functions to be performed.

For example, to make a machine that can be instructed in natural language, it is necessary to probe deeply into the nature of language. In order to make a machine capable of learning, we have to probe deeply into the nature of learning. And from this kind of research comes the broader definition of artificial intelligence: that of a cognitive science. In this sense, AI shares its domain with the older disciplines such as linguistics and psychology. But what is distinctive in AI is that its methodology and style of theorizing draw heavily on theories of computation. In this chapter we shall use this style of theorizing in several ways: first, to reinterpret Piaget; second, to develop the theories of learning and understanding that inform our design of educational situations; and third, in a somewhat more unusual way. The aim of AI is to give concrete form to ideas

157

about thinking that previously might have seemed abstract, even metaphysical. It is this concretizing quality that has made ideas from AI so attractive to many contemporary psychologists. We propose to teach AI to children so that they, too, can think more concretely about mental processes. While psychologists use ideas from AI to build formal, scientific theories about mental processes, children use the same ideas in a more informal and personal way to think about themselves. And obviously I believe this to be a good thing in that the ability to articulate the processes of thinking enables us to improve them.

Piaget has described himself as an epistemologist. What does he mean by that? When he talks about the developing child, he is really talking as much about the development of knowledge. This statement leads us to a contrast between epistemological and psychological ways of understanding learning. In the psychological perspective, the focus is on the laws that govern the learner rather than on what is being learned. Behaviorists study reinforcement schedules, motivation theorists study drive, gestalt theorists study good form. For Piaget, the separation between the learning process and what is being learned is a mistake. To understand how a child learns number, we have to study number. And we have to study number in a particular way: We have to study the structure of number, a mathematically serious undertaking. This is why it is not at all unusual to find Piaget referring in one and the same paragraph to the behavior of small children and to the concerns of theoretical mathematicians. To make more concrete the idea of studying learning by focusing on the structure of what is learned, we look at a very concrete piece of learning from everyday life and see how different it appears from a psychological and from an epistemological perspective.

We will consider learning to ride a bicycle. If we did not know better riding a bicycle would seem to be a really remarkable thing. What makes it possible? One could pursue this question by studying the rider to find out what special attributes (speed of reaction, complexity of brain functioning, intensity of motivation) contribute to his performance. This inquiry, interesting though it might be, is irrelevant to the real solution to the problem. People can ride bicycles because the bicycle, once in motion, is inherently stable. A bicycle without a rider pushed off on a steep downgrade will not fall over; it will run indefinitely down the hill. The geometrical construction of the front fork ensures that if the bicycle leans to the left the wheel will rotate to the left, thus causing that bicycle to turn and produce a centrifugal force that throws the bicycle to the right, counteracting the tendency to fall. The bicycle without a rider balances perfectly well. With a novice rider it will fall. This is because the novice has the wrong intuitions about balancing and freezes the position of the bicycle so that its own corrective mechanism cannot work freely. Thus learning to ride does not mean learning to balance, it means learning not to unbalance, learning not to interfere.

What we have done here is understand a process of learning by acquiring deeper insight into what was being learned. Psychological principles had nothing to do with it. And just as we have understood how people ride bicycles by studying bicycles, Piaget has taught us that we should understand how children learn number through a deeper understanding of what number is.

Mathematicians interested in the nature of number have looked at the problem from different standpoints. One approach, associated with the formalists, seeks to understand number by setting up axioms to capture it. A second approach, associated with Bertrand Russell, seeks to define number by reducing it to something more fundamental, for example, logic and set theory. Although both of these approaches are valid, important chapters in the history of mathematics, neither casts light on the question of why number is learnable. But there is a school of mathematics that does do so, although this was not its intention. This is the structuralism of the Bourbaki school.[1] Bourbaki is a pseudonym taken by a group of French mathematicians who set out to articulate a uniform theory for mathematics. Mathematics was to be one, not a collection of subdisciplines each with its own language and line of development. The school moved in this direction by recognizing a number of building blocks that it called the "mother structures." These structures have something in common with our idea of microworlds. Imagine a microworld in which things can be ordered but have no other properties. The knowledge of how to work the world is, in terms of the Bourbaki school, the mother structure of order. A second microworld allows relations of proximity, and this is the mother structure of topology. A third has to do with combining entities to produce new entities; this is the algebraic microstructure. The Bourbaki school's unification of mathematics is achieved by seeing more complex structures, such as arithmetic, as combinations of simpler structures of which the most important are the three mother structures. This school had no intention of making a theory of learning. They intended their structural analysis to be a technical tool for mathematicians to use in their day-to-day work. But the theory of mother structures *is* a theory of learning. It is a theory of how number is learnable. By showing how the structure of arithmetic can be decomposed into simpler, but still meaningful and coherent, structures, the mathematicians are showing a mathetic pathway into numerical knowledge. It is not surprising that Piaget, who was explicitly searching for a theory of number that would explain its development in children, developed a similar, parallel set of constructs, and then, upon "discovering" the Bourbaki school was able to use its constructs to elaborate his own.

Piaget observed that children develop coherent intellectual structures that seemed to correspond very closely to the Bourbaki mother structures. For example, recall the Bourbaki structure of order; indeed, from the earliest ages, children begin to develop expertise in ordering things. The topological and algebraic mother structures have similar developmental precursors. What makes them learnable? First of all, each represents a coherent activity in the child's life that could in principle be learned and made sense of independent of the others.

Second, the knowledge structure of each has a kind of internal simplicity that Piaget has elaborated in his theory of *groupements*, and which will be discussed in slightly different terms later. Third, although these mother structures are independent, the fact that they are learned in parallel and that they share a common formalism are clues that they are mutually supportive; the learning of each facilitates the learning of the others.

Piaget has used these ideas to give an account of the development of a variety of domains of knowledge in terms of a coherent, lawful set of structures as processes within the child's mind. He describes these internal structures as always in interaction with the external world, but his theoretical emphasis has been the internal events. My perspective is more interventionist. My goals are education, not just understanding. So, in my own thinking I have placed a greater emphasis on two dimensions implicit but not elaborated in Piaget's own work: an interest in intellectual structures that could develop as opposed to those that actually at present do develop in the child, and the design of learning environments that are resonant with them. The Turtle can be used to illustrate both of these interests: first, the identification of a powerful set of mathematical ideas that we do not presume to be represented, at least not in a developed form, in children; second, the creation of a *transitional object,* the Turtle, that can exist in the child's environment and make contact with the ideas. As a mathematician I know that one of the most powerful ideas in the history of science was that of differential analysis. From Newton onward, the relationship between the local and the global pretty well set the agenda for mathematics. Yet this idea has had no place in the world of children, largely because traditional access to it depends on an infrastructure of formal, mathematical training. For most people, nothing is more natural than that the most advanced ideas in mathematics should be inaccessible to children. From the perspective I took from Piaget, we would expect to find connections. So we set out to find some. But finding the connections did not simply mean inventing a new kind of clever, "motivating" pedagogy. It meant a research agenda that included separating what was most powerful in the idea of *differential* from the accidents of inaccessible formalisms. The goal was then to connect these scientifically fundamental structures with psychologically powerful ones. And of course these were the ideas that underlay the Turtle circle, the physics microworlds, and the touch-sensor Turtle.

In what sense is the natural environment a source of microworlds, indeed a source for a network of microworlds? Let's narrow the whole natural environment to those things in it that may serve as a source for one specific microworld, a microworld of pairing, of one-to-one correspondence. Much of what children see comes in pairs: mothers and fathers, knives and forks, eggs and egg cups. And they, too, are asked to be active constructors of pairs. They are asked to sort socks, lay the table with one place setting for each person, and distribute candies. When children focus attention on pairs they are in a self-constructed microworld, a microworld of pairs, in the same sense as we placed our students in the microworlds of geometry and physics Turtles. In both cases the relevant microworld is stripped of complexity, is simple, graspable. In both cases the child is allowed to play freely with its elements. Although there are constraints on the materials, there are no constraints on the exploration of combinations. And in both cases the power of the environment is that it is "discovery rich."

Working with computers can make it more apparent that children construct their own personal microworlds. The story of Deborah at the end of chapter 4 is a good example. LOGO gave her the opportunity to construct a particularly tidy microworld, her "RIGHT 30 world." But she might have done something like this in her head without a computer. For example, she might have decided to understand directions in the real world in terms of a simple set of operations. Such intellectual events are not usually visible to observers, any more than my algebra teachers knew that I used gears to think about equations. But they can be seen if one looks closely enough. Robert Lawler, a member of the Massachusetts Institute of Technology LOGO group, demonstrated this most clearly in his doctoral research. Lawler set out to observe everything a six-year-old child, his daughter Miriam, did during a six-month period. The wealth of information he obtained allowed him to piece together a picture of the microstructure of Miriam's growing abilities. For example, during this period Miriam learned to add, and Lawler was able to show that this did not consist of acquiring one logically uniform procedure. A better model of her learning to add is that she brought into a working relationship a number of idiosyncratic microworlds, each of which could be traced to identifiable, previous experiences.

I have said that Piaget is an epistemologist, but have not elaborated on what kind. Epistemology is the theory of knowledge. The term epistemology could, according to its etymology, be used to cover all knowledge about knowledge, but traditionally it has been used in a rather special way; that is, to describe the study of the conditions of validity of knowledge. Piaget's epistemology is concerned not with the validity of knowledge but with its origin and growth. He is concerned with the genesis and evolution of knowledge, and marks this fact by describing his field of study as "genetic epistemology." Traditional epistemology has often been taken as a branch of philosophy. Genetic epistemology works to assert itself as a science. Its students gather data and develop theories about how knowledge developed, sometimes focusing on the evolution of knowledge in history, sometimes on the evolution of knowledge in the individual. But it does not see the two realms as distinct: It seeks to understand relations between them. These relations can take different forms.

In the simplest case the individual development is parallel to the historical development, recalling the biologists' dictum, ontogeny recapitulates phylogeny. For example, children uniformly represent the physical world in an Aristotelian manner, thinking, for example, that forces act on position rather than on velocities. In other cases, the relation is more complex, indeed to the point of reversal. Intellectual structures that appear first in a child's development are sometimes characteristic not of early science but of the most modern science. So, for example, the mother structure topology appears very early in the child's development, but topology itself appeared as a mathematical subdiscipline only in modern times. Only when mathematics becomes sufficiently advanced is it able to discover its own origins.

In the early part of the twentieth century, formal logic was seen as synonymous with the foundation of mathematics. Not until Bourbaki's structuralist theory appeared do we see an internal development in mathematics that opened the field up to "remembering" its genetic roots. And through the work of genetic epistemology, this "remembering" puts mathematics in the closest possible relationship to the development of research about how children construct their reality.

Genetic epistemology has come to assert a set of homologies between the structures of knowledge and the structures of the mind that come into being to grasp this knowledge. Bourbaki's mother structures are not simply the elements that underly the concept of number; rather, homologies are found in the mind as it constructs number for itself. Thus, the importance of studying the structure of knowledge is not just to better understand the knowledge itself, but to understand the person.

Research on the structure of this dialectical process translates into the belief that neither people nor knowledge—including mathematics—can be fully grasped separately from the other, a belief that was eloquently expressed by Warren McCulloch, who, together with Norbert Wiener, should have credit for founding cybernetics. When asked, as a youth, what question would guide his scientific life, McCulloch answered: "What is a man so made that he can understand number and what is number so made that a man can understand it?"

For McCulloch as for Piaget, the study of people and the study of what they learn and think are inseparable. Perhaps paradoxically for some, research on the nature of that inseparable relationship has been advanced by the study of machines and the knowledge they can embody. And it is to this research methodology, that of artificial intelligence, that we now turn.

In artificial intelligence, researchers use computational models to gain insight into human psychology as well as reflect on human psychology as a source of ideas about how to make mechanisms emulate human intelligence. This enterprise strikes many as illogical: Even when the performance looks identical, is there any reason to think that underlying processes are the same? Others find it illicit: The line between man and machine is seen as immutable by both theology and mythology. There is a fear that we will dehumanize what is essentially human by inappropriate analogies between our "judgments" and those computer "calculations." I take these objections very seriously, but feel that they are based on a view of artificial intelligence that is more reductionist that anything I myself am interested in. A brief parable and some only half-humored reasoning by analogy express my own views on the matter.

Men have always been interested in flying. Once upon a time, scientists determined to understand how birds fly. First they watched them, hoping to correlate the motion of a bird's wings with its upward movement. Then they proceeded to experiment and found that when its feathers were plucked, a bird could no longer fly. Having thus determined that feathers were the organ of flight, the scientists then focused their efforts on microscopic and ultramicroscopic investigation of feathers in order to discover the nature of their flight-giving power.

In reality our current understanding of how birds fly did not come through a study narrowly focused on birds and gained nothing at all from the study of feathers. Rather, it came from studying phenomena of different kinds and requiring different methodologies. Some research involved highly mathematical studies in the laws of motion of idealized fluids. Other research, closest to our central point here, consisted of building machines for "artificial flight." And, of course, we must add to the list the actual observation of bird flight. All these research activities synergistically gave rise to aeronautical science through what we understand of the "natural flight" of birds and the "artificial flight" of airplanes. And it is in much the same spirit that I imagine diverse investigations in mathematics and in machine intelligence to act synergistically with psychology in giving rise to a discipline of cognitive science whose principles would apply to natural and to artificial intelligence.

It is instructive to transpose to the context of flying the common objections raised against AI. This leads us to imagine skeptics who would say, "You mathematicians deal with idealized fluids—the real atmosphere is vastly more complicated," or "You have no reason to suppose that airplanes and birds work the same way—birds have no propellors, airplanes have no feathers." But the premises of these criticisms are true only in the most superficial sense: the same *principle* (e.g., Bernoulli's law) applies to real as well as ideal fluids, and they apply whether the fluid flows over a feather or an aluminum wing.

Workers in the "cognitive studies" branch of AI do not share any one way of thinking about thinking, any more than traditional psychologists do. For some, the computer model is used to reduce all thinking to the formal operations of powerful deductive systems.

165

Aristotle succeeded in formulating the deductive rules for a small corner of human thinking in such simple syllogisms as "If all men are mortal and Socrates is a man, then Socrates is mortal." In the nineteenth century, mathematicians were able to extend this kind of reasoning to a somewhat larger but still restricted area. But only in the context of computational methods has there been a serious attempt to extend deductive logic to cover all forms of reasoning, including common-sense reasoning and reasoning by analogy. Working with this kind of deductive model was very popular in the early days of AI. In recent years, however, many workers in the field have adopted an almost diametrically opposed strategy. Instead of seeking powerful deductive methods that would enable surprising conclusions to be drawn from general principles, the new approach assumes that people are able to think only because they can draw on larger pools of specific, particular knowledge. More often than we realize, we solve problems by "almost knowing the answer" already. Some researchers try to make programs be intelligent by giving them such quantities of knowledge that the greater part of solving a problem becomes its retrieval from somewhere in the memory.

Given my background as a mathematician and Piagetian psychologist, I naturally became most interested in the kinds of computational models that might lead me to better thinking about powerful developmental processes: the acquisition of spatial thinking and the ability to deal with size and quantity. The rival approaches—deductive and knowledge based—tended to address performance of a given intellectual system whose structure, if not whose content, remained static. The kind of developmental questions I was interested in needed a dynamic model for how intellectual structures themselves could come into being and change. I believe that these are the kind of models that are most relevant to education.

The best way I know to characterize this approach is to give a sample of a theory heavily influenced by ideas from computation that can help us understand a specific psychological phenomenon: Piagetian conservation. We recall that children up until the age of six or seven believe that a quantity of liquid can increase or decrease when it is poured from one container to another. Specifically, when the second container is taller and narrower than the first, the children unanimously assert that the quantity of liquid has increased. And then, as if by magic, at about the same age, all children change their mind: They now just as unequivocally insist that the amount of liquid remains the same.

Many theories have been advanced for how this could come to pass. One of them, which may sound most familiar because it draws on traditional psychological categories, attributes the pre-conservationist position to the child's being dominated by "appearances." The child's "reason" cannot override how things "seem to be." Perception rules.

Let us now turn to another theory, this time one inspired by computational methods. Again we ask the question: Why does height in a narrow vessel seem like more to the child, and how does this change?

Let us posit the existence of three agents in the child's mind, each of which judges quantities in a different "simple-minded" way.* The first, A_{height} judges the quantity of liquids and of anything else by its vertical extent. A_{height} is a practical agent in the life of the child. It is accustomed to comparing children by standing them back to back and of equalizing the quantities of Coca-Cola and chocolate milk in children's glasses. We emphasize that A_{height} does not do anything as complicated as "perceive" the quantity of liquid. Rather, it is fanatically dedicated to an abstract principle: Anything that goes higher is more.

There is a second agent, called A_{width}, that judges by the horizontal extent. It is not as "practiced" as A_{height}. It gets its chance to judge that there is a lot of water in the sea, but in the mind of the child this principle is less "influential" than A_{height}.

Finally, there is an agent called $A_{history}$ that says that the quantities are the same because once they were the same. $A_{history}$ seems to speak like a conservationist child, but this is an illusion. $A_{history}$ has

* The computational perspective on conservation that follows is a highly schematized and simplified overview of how this phenomenon would be explained by a theory, "The Society of Mind," being developed by Marvin Minsky and the author and to be discussed in our forthcoming book.

no understanding and would say the quantity is the same even if some had indeed been added.

In the experiment with the preconservationist child, each of the three agents makes its own "decision" and clamors for it to be adopted. As we know, A_{height}'s voice speaks the loudest. But this changes as the child moves on to the next stage.

There are three ways, given our assumption of the presence of agents, for this change to take place. A_{height} and A_{width} could become more "sophisticated," so that, for example, A_{height} would disqualify itself except when all other things are equal. This would mean that A_{height} would only step forward to judge by height those things that have equal cross sections. Second, there could be a change in "seniority," in prerogative: $A_{history}$ could become the dominant voice. Neither of these two modes of change is impossible. But there is a third mode that produces the same effect in a simpler way. Its key idea is that A_{height} and A_{width} neutralize one another by giving contradictory opinions. The idea is attractive (and close to Piaget's own concept of grouplike compositions of operations) but raises some problems. Why do all three agents not neutralize one another so that the child would have no opinion at all? The question is answered by a further postulate (which has much in common with Piaget's idea that intellectual operators be organized into *groupements*). The principle of neutralization becomes workable if enough structure is imposed on the agents for A_{height} and A_{width} to be in a special relationship with one another but not with $A_{history}$. We have seen that the technique of creating a new entity works powerfully in programming systems. And this is the process we postulate here. A new entity, a new agent comes into being. This is A_{geom}, which acts as the supervisor for A_{height} and A_{width}. In cases where A_{height} and A_{width} agree, A_{geom} passes on their message with great "authority." But if they disagree, A_{geom} is undermined and the voices of the underlings are neutralized. It must be emphasized that A_{geom} is not meant to "understand" the reasons for decision making by A_{height} and A_{width}. A_{geom} knows nothing except whether they agree and, if so, in which direction.

This model is absurdly oversimplified in suggesting that even so simple a piece of a child's thinking (such as this conservation) can be understood in terms of interactions of four agents. Dozens or hundreds are needed to account for the complexity of the real process. But, despite its simplicity, the model accurately conveys some of the principles of the theory: in particular, that the components of the system are more like people than they are like propositions and their interactions are more like social interactions than like the operations of mathematical logic. This shift in perspective allows us to solve many technical problems in developmental psychology. In particular, we can understand logical learning as continuous with social and bodily learning.

I have said that this theory is inspired by a computational metaphor. One might ask how. The "theory" might appear to be nothing but anthropomorphic talk. But we have already seen that anthropomorphic descriptions are often a step toward computational theories. And the thrust of the society-of-mind theory is that agents can be translated into precise computational models. As long as we only think about these agents as "people," the theory is circular. It explains the behavior of people in terms of the behavior of people. But, if we can think of the agents as well-defined computational entities similar to the subprocedures VEE, LINE, and HEAD in the procedure MAN, everything becomes clearer. We saw even in small programs how very simple modules can be put together to produce complex results.

This computational argument saves the society-of-mind theory from the charge of relying on a vicious circle. But it does not save it from being circular: On the contrary, like recursive programs in the style of the procedure SPI of chapter 3, the theory derives much of its power from a constructive use of "circular logic." A traditional logician looking at how SPI was defined by reference to SPI might have objected, but the computer programmers and genetic epistemologists share a vision in which this kind of self-reference is not only legitimate but necessary. And both see it as having an element of paradox that is only very partially captured by noting how children use their "inferior" logic to construct the "superior" logic of their next phase of development. To an increasing extent throughout his long career Piaget has emphasized the importance for intellectual growth of children's ability to reflect on their own thinking.

169

The "mathetic paradox" lies in the fact that this reflection must be from within the child's current intellectual system.

Despite its oversimplified, almost metaphorical status, the four-agent account of conservation captures an element of the paradox. A mathematical logician might like to impose on A_{height} and A_{width} a superior agent capable of calculating, or at least estimating, volume from height and cross-section. Many educators might like to impose such a formula on the child. But this would be introducing an element alien to the pre-conservationist child's intellectual system. Our A_{geom} belongs firmly in the child's system. It might even be derived from the model of a father not quite succeeding in imposing order on the family. It is possible to speculate, though I have no evidence, that the emergence of conservation is related to the child's oedipal crisis through the salience it gives to this model. I feel on firmer ground in guessing that something like A_{geom} can become important because it so strongly has the two-sided relationship that was used to conceive the Turtle: It is related both to structures that are firmly in place, such as the child's representation of authority figures, and to germs of important mathematical ideas, such as the idea of "cancellation."

Readers who are familiar with Piaget's technical writings will recognize this concept germ as one of the principles in his "groupments." They may therefore see our model as not very different from Piaget's. In a fundamental sense they would be right. But a new element is introduced in giving a special role to computational structures: The theme of this book has been the idea of exploiting this special role by giving children access to computational cultures. If, and only if, these have the right structure they may greatly enhance children's ability to represent the structures-in-place in ways that will mobilize their conceptual potential.

To recapitulate our reinterpretation of Piaget's theory makes three points. First, it provides a specific psychological theory, highly competitive in its parsimony and explanatory power with others in the field. Second, it shows us the power of a specific computational principle, in this case the theory of pure procedures, that is, procedures that can be closed off and used in a modular way. Third, it concretizes my argument about how different languages can influence the cultures that can grow up around them. Not all programming languages embody this theory of pure procedures. When they do not, their role as metaphors for psychological issues is severely biased. The analogy between artificial intelligence and artificial flying made the point that the same principles could underlie the artificial and natural phenomena, however different these phenomena might appear. The dynamics of lift are fundamental to flight as such, whether the flyers are of flesh and blood or of metal. We have just seen a principle that may be fundamental both to human and artificial intelligence: the principle of epistemological modularity. There have been many arguments about whether the ideal machine for the achievement of intelligence would be analog or digital, and about whether the brain is analog or digital. From the point of view of the theory I am advancing here, these arguments are beside the point. The important question is not whether the brain or the computer is discrete but whether knowledge is modularizable.

For me, our ability to use computational metaphors in this way, as carriers for new psychological theories, has implications concerning where theories of knowledge are going and where we are going as producers and carriers of knowledge. These areas are not independent. In earlier chapters it was suggested that how we think about knowledge affects how we think about ourselves. In particular, our image of knowledge as divided up into different kinds leads us to a view of people as divided up according to what their aptitudes are. This in turn leads to a balkanization of our culture.

Perhaps the fact that I have spoken so negatively about the balkanization of our culture and so positively about the modularization of knowledge requires some clarification. When knowledge can be broken up into "mind-size bites," it is more communicable, more assimilable, more simply constructable. The fact that we divide knowledge up into scientific and humanistic worlds defines some knowledge as being a priori uncommunicable to certain kinds of people. Our commitment to communication is not only expressed through our commitment to modularization, which facilitates it, but through our attempt to find a language for such domains as physics and mathematics, which have as their essence communication between constructed entities. By restating Newton's laws as assertions about how particles (or "Newtonian Turtles") communicate with one another, we give it a handle that can be more easily grabbed by a child or by a poet.

Consider another example of how our images of knowledge can subvert our sense of ourselves as intellectual agents. Educators sometimes hold up an ideal of knowledge as having the kind of coherence defined by formal logic. But these ideals bear little resemblance to the way in which most people experience themselves. The subjective experience of knowledge is more similar to the chaos and controversy of competing agents than to the certitude and orderliness of p's implying q's. The discrepancy between our experience of ourselves and our idealizations of knowledge has an effect: It intimidates us, it lessens the sense of our own competence, and it leads us into counterproductive strategies for learning and thinking.

Many older students have been intimidated to the point of dropping out, and what is true for adults is doubly true for children. We have already seen that despite their experience of themselves as theory builders, children are not respected as such. And these contradictions are compounded by holding out an ideal of knowledge to which no one's thinking conforms. Many children and college students who decide "I can never be a mathematician or a scientist" are reflecting a discrepancy between the way they are led to believe the mathematician must think and the way they know they do. In fact the truth is otherwise: Their own thinking is much more like the mathematician's than either is like the logical ideal.

I have spoken of the importance of powerful ideas in grasping the world. But we could hardly ever learn a new idea if every time we did we had to totally reorganize our cognitive structures in order to use it or if we even had to insure that no inconsistencies had been introduced. Although powerful ideas have the capacity to help us organize our way of thinking about a particular class of problems (such as physics problems), we *don't* have to reorganize ourselves in order to use them. We put our skills and heuristic strategies into a kind of tool box—and while their interaction can, in the course of time, give rise to global changes, the act of learning is itself a local event.

The local nature of learning is seen in my description of the acquisition of conservation. The necessary agents entered the system locally; their top goals were in contradiction with each other; the agent that finally reconciles them leaves them in place. There is no reason why this "patchwork theory" of theory building should be considered appropriate only for describing the learning of children. Research in artificial intelligence is gradually giving us a surer sense of the range of problems that can be meaningfully solved on the pattern we have sketched for the conservation problem: with modular agents, each of them simple-minded in its own way, many of them in conflict with one other. The conflicts are regulated and kept in check rather than "resolved" through the intervention of special agents no less simple-minded than the original ones. Their way of reconciling differences does not involve forcing the system into a logically consistent mold.

The process reminds one of tinkering; learning consists of building up a set of materials and tools that one can handle and manipulate. Perhaps most central of all, it is a process of working with what you've got. We're all familiar with this process on the conscious level, for example, when we attack a problem empirically, trying out all the things that we have ever known to have worked on similar problems before. But here I suggest that working with what you've got is a shorthand for deeper, even unconscious learning processes. Anthropologist Claude Lévi-Strauss[2] has spoken in similar terms of the kind of theory building that is characteristic of primitive science. This is a science of the concrete, where the relationships between natural objects in all their combinations and recombinations provide a conceptual vocabulary for building scientific theories. Here I am suggesting that in the most fundamental sense, we, as learners, are all *bricoleurs*.[3] This leads us into the second kind of implication of our computational theory of agents. If the first implications had to do with impacts on our ideas about knowledge and learning, the second have to do with possible impacts on our images of ourselves as learners. If *bricolage* is a model for how scientifically legitimate theories are built, then we can begin to develop a greater respect for ourselves as *bricoleurs*. And of course this joins our central theme of the importance and power of Piagetian learning. In order to create the conditions for bringing what is now non-Piagetian learning to the Piagetian side, we have to be able to act in good faith. We have to feel that we are not denaturing knowledge in the process.

I end this chapter on cognitive theory and people with a conjecture. Earlier I said that I would not present Piaget as a theorist of stages. But thinking about Piagetian stages does provide a context in which to make an important point about a possible impact of a computational culture on people. Piaget sees his stages of cognitive development as invariable, and numerous cross-cultural investigations have seemed to confirm the validity of his belief. In society after society, children seem to develop cognitive capacities in the same order. In particular, his stage of concrete operations, to which the conservations typically belong, begins four or more years earlier than the next and final stage, the stage of formal operations. The construct of a stage of concrete operations is supported by the observation that, typically, children in our society at six or seven make a breakthrough in many realms, and seemingly all at once. They are able to use units of numbers, space, and time; to reason by transitivity; to build up classificatory systems. But there are things they cannot do. In particular, they flounder in situations that call for thinking not about how things are but about all the ways they could be. Let us consider the following example, which I anticipated in the introduction.

A child is given a collection of beads of different colors, say green, red, blue, and black, and is asked to construct all the possible pairs of colors: green-blue, green-red, green-black, and then the triplets and so on. Just as children do not acquire conservation until their seventh year, children around the world are unable to carry out such combinatorial tasks before their eleventh or twelfth year. Indeed, many adults who are "intelligent" enough to live normal lives never acquire this ability.

What is the nature of the difference between the so-called "concrete" operations involved in conservation and the so-called "formal" operations involved in the combinatorial task? The names given them by Piaget and the empirical data suggest a deep and essential difference. But looking at the problem through the prism of the ideas developed here gives a much different impression.

From a computational point of view, the most salient ingredients of the combinatorial task are related to the idea of procedure—systematicity and debugging. A successful solution consists of following some such procedure as:

1. Separate the beads into colors.
2. Choose a color A as color 1.
3. Form all the pairs that can be formed with color 1.
4. Choose color 2.
5. Form all the pairs that can be formed with color 2.
6. Do this for each color.
7. Go back and remove the duplicates.

So what is really involved is writing and executing a program including the all-important debugging step. This observation suggests a reason for the fact that children acquire this ability so late: Contemporary culture provides relatively little opportunity for *bricolage* with the elements of systematic procedures of this type. I do not mean to say that there are no such opportunities. Some are encountered; for example, in games where a child can create his own "combinatorial microworlds." But the opportunities, the incentives, and the help offered the child in this area are very significantly less than in such areas as number. In our culture number is richly represented, systematic procedure is poorly represented.[4]

I see no reason to doubt that this difference could account for a gap of five years or more between the ages at which conservation of number and combinatorial abilities are acquired.

The standard methodology for investigating such hypotheses as this is to compare children in different cultures. This has, of course, been done for the Piagetian stages. Children at all the levels of development anthropologists have been able to distinguish, and in over a hundred different societies from all the continents, have been asked to pour liquids and sort beads. In all cases, if conservation and combinatorial skills came at all, conservation of numbers was evidenced by children five or more years younger than those evidencing combinatorial skills. Yet this observation casts no doubt on my hypothesis. It may well be universally true of precomputer societies that *numerical* knowledge would be more richly represented than *programming* knowledge. It is not hard to invent plausible explanations of such a cognitive-social universal. But things may be different in the computer-rich cultures of the future. If computers and programming become a part of the daily life of children, the conservation-combinatorial gap will surely close and could conceivably be reversed: Children may learn to be systematic before they learn to be quantitative!

Chapter 8

Images of the Learning Society

THE VISION I HAVE PRESENTED is of a particular computer culture, a mathetic one, that is, one that helps us not only to learn but to learn about learning. I have shown how this culture can humanize learning by permitting more personal, less alienating relationships with knowledge and have given some examples of how it can improve relationships with other people encountered in the learning process: fellow students and teachers. But I have made only passing remarks about the social context in which this learning might take place. It is time to face (though I cannot answer) a question that must be in many readers' minds: Will this context be school?

The suggestion that there might come a day when schools no longer exist elicits strong response from many people. There are many obstacles to thinking clearly about a world without schools. Some are highly personal. Most of us spent a larger fraction of our lives going to school than we care to think about. For example, I am over fifty and yet the number of my postschool years has barely caught up with my preschool and school years. The concept of a world without school is highly dissonant with out experiences of our own lives. Other obstacles are more conceptual. One cannot define such a world negatively, that is by simply removing school and putting nothing in its place. Doing so leaves a thought vacuum that the mind has to fill one way or another, often with vague but scary images of children "running wild," "drugging themselves," or "making life impossible for their parents." Thinking seriously about a world without schools calls for elaborated models of the nonschool activities in which children would engage.

For me, collecting such models has become an important part of thinking about the future of children. I recently found an excellent model during a summer spent in Brazil. For example, at the core of the famous carnival in Rio de Janeiro is a twelve-hour-long procession of song, dance, and street theater. One troop of players after another presents its piece. Usually the piece is a dramatization through music and dance of a historical event or folk tale. The lyrics, the choreography, the costumes are new and original. The level of technical achievement is professional, the effect breathtaking. Although the reference may be mythological, the processions are charged with contemporary political meaning.

The processions are not spontaneous. Preparing them as well as performing in them are important parts of Brazilian life. Each group prepares separately—and competitively—in its own learning environment, which is called a samba school. These are not schools as we know them; they are social clubs with memberships that may range from a few hundred to many thousands. Each club owns a building, a place for dancing and getting together. Members of a samba school go there most weekend evenings to dance, to drink, to meet their friends.

During the year each samba school chooses its theme for the next carnival, the stars are selected, the lyrics are written and rewritten, the dance is choreographed and practiced. Members of the school range in age from children to grandparents and in ability from novice to professional. But they dance together and as they dance everyone is learning and teaching as well as dancing. Even the stars are there to learn their difficult parts.

Every American disco is a place for learning as well as for dancing. But the samba schools are very different. There is a greater social cohesion, a sense of belonging to a group, and a sense of common purpose. Much of the teaching, although it takes place in a natural environment, is deliberate. For example, an expert dancer gathers a group of children around. For five or for twenty minutes a specific learning group comes into existence. Its learning is deliberate and focused. Then it dissolves into the crowd.

In this book we have considered how mathematics might be learned in settings that resemble the Brazilian samba school, in settings that are real, socially cohesive, and where experts and novices are all learning. The samba school, although not "exportable" to an alien culture, represents a set of attributes a learning environment should and could have. Learning is not separate from reality. The samba school has a purpose, and learning is integrated in the school for this purpose. Novice is not separated from expert, and the experts are also learning.

LOGO environments are like samba schools in some ways, unlike them in other ways. The deepest resemblance comes from the fact that in them mathematics is a real activity that can be shared by novices and experts. The activity is so varied, so discovery-rich, that even in the first day of programming, the student may do something that is new and exciting to the teacher. John Dewey expressed a nostalgia for earlier societies where the child becomes a hunter by real participation and by playful imitation. Learning in our schools today is not significantly participatory—and doing sums is not an imitation of an exciting, recognizable activity of adult life. But writing programs for computer graphics or music and flying a simulated spaceship do share very much with the real activities of adults, even with the kind of adult who could be a hero and a role model for an ambitious child.

LOGO environments also resemble samba schools in the quality of their human relationships. Although teachers are usually present, their interventions are more similar to those of the expert dancers in the samba school than those of the traditional teacher armed with lesson plans and a set curriculum. The LOGO teacher will answer questions, provide help if asked, and sometimes sit down next to a student and say: "Let me show you something." What is shown is not dictated by a set syllabus. Sometimes it is something the student can use for an immediate project. Sometimes it is something that the teacher has recently learned and thinks the student would enjoy. Sometimes the teacher is simply acting spontaneously as people do in all unstructured social situations when they are excited about what they are doing. The LOGO environment is like the samba school also in the fact that the flow of ideas and even of instructions is not a one-way street. The environment is designed to foster richer and deeper interactions than are commonly seen in schools today in connection with anything mathematical. Children create programs that produce pleasing graphics, funny pictures, sound effects, music, and computer jokes. They start interacting mathematically because the product of their mathemetical work belongs to them and belongs to real life. Part of the fun is sharing, posting graphics on the walls, modifying and experimenting with each other's work, and bringing the "new" products back to the original inventors. Although the work at the computer is usually private it increases the children's desire for interaction. These children want to get together with others engaged in similar activities because they have a lot to talk about. And what they have to say to one another is not limited to talking about their products: LOGO is designed to make it easy to tell about the process of making them.

By building LOGO in such a way that structured thinking becomes powerful thinking, we convey a cognitive style, one aspect of which is to facilitate talking about the process of thinking. LOGO's emphasis on debugging goes in the same direction. Students' bugs become topics of conversation; as a result they develop an articulate and focused language to use in asking for help when it is needed. And when the need for help can be articulated clearly, the helper does not necessarily have to be a specially trained professional in order to give it. In this way the LOGO culture enriches and facilitates the interaction between all participants and offers opportunities for more articulate, effective, and honest teaching relationships. It is a step toward a situation in which the line between learners and teachers can fade.

Despite these similarities, LOGO environments are not samba schools. The differences are quite fundamental. They are reflected superficially in the fact that the teachers are professionals and are in charge even when they refrain from exerting authority. The students are a transitory population and seldom stay long enough to make LOGO's long-term goals their own. Ultimately the difference has to do with how the two entities are related to the surrounding culture. The samba school has rich connections with a popular culture. The knowledge being learned there is continuous with that culture. The LOGO environments are artificially maintained oases where people encounter knowledge (mathematical and mathetic) that has been separated from the mainstream of the surrounding culture, indeed which is even in some opposition to values expressed in that surrounding culture. When I ask myself whether this can change, I remind myself of the social nature of the question by remembering that the samba school was not designed by researchers, funded by grants, nor implemented by government action. It was not made. It happened. This must be true too of any new successful forms of associations for learning that might emerge out of the mathetic computer culture. Powerful new social forms must have their roots in the culture, not be the creatures of bureaucrats.

Thus we are brought back to seeing the necessity for the educator to be an anthropologist. Educational innovators must be aware that in order to be successful they must be sensitive to what is happening in the surrounding culture and use dynamic cultural trends as a medium to carry their educational interventions.

It has become commonplace to say that today's culture is marked by a ubiquitous computer technology. This has been true for some time. But in recent years, there *is* something new. In the past two years, over 200,000 personal computers have entered the lives of Americans, some of them originally bought for business rather than recreational or educational purposes. What is important to the educator-as-anthropologist, however, is that they exist as objects that people see, and start to accept, as part of the reality of everyday life. And at the same time that this massive penetration of the technology is taking place, there is a social movement afoot with great relevance for the politics of education. There is an increasing disillusion with traditional education. Some people express this by extreme action, actually withdrawing their children from schools and choosing to educate them at home. For most, there is simply the gnawing sense that schools simply aren't doing the job anymore. I believe that these two trends can come together

MINDSTORMS

in a way that would be good for children, for parents, and for learning. This is through the construction of educationally powerful computational environments that will provide alternatives to traditional classrooms and traditional instruction. I do not present LOGO environments as my proposal for this. They are too primitive, too limited by the technology of the 1970s. The role I hope they fill is that of a model. By now the reader must anticipate that I shall say an object-to-think-with, that will contribute to the essentially social process of constructing the education of the future.

LOGO environments are not samba schools, but they are useful for imagining what it would be like to have a "samba school for mathematics." Such a thing was simply not conceivable until very recently. The computer brings it into the realm of the possible by providing mathematically rich activities which could, in principle, be truly engaging for novice and expert, young and old. I have no doubt that in the next few years we shall see the formation of some computational environments that deserve to be called "samba schools for computation." There have already been attempts in this direction by people engaged in computer hobbyist clubs and in running computer "drop-in centers."

In most cases, although the experiments have been interesting and exciting, they have failed to make it because they were too primitive. Their computers simply did not have the power needed for the most engaging and shareable kinds of activities. Their visions of how to integrate computational thinking into everyday life was insufficiently developed. But there will be more tries, and more and more. And eventually, somewhere, all the pieces will come together and it will "catch." One can be confident of this because such attempts will not be isolated experiments operated by researchers who may run out of funds or simply become disillusioned and quit. They will be manifestations of a social movement of people interested in personal computation, interested in their own children, and interested in education.

There are problems with the image of samba schools as the locus of education. I am sure that a computational samba school will catch on somewhere. But the first one will almost certainly happen in a community of a particular kind, probably one with a high density of middle-income engineers. This will allow the computer samba school to put down "cultural roots," but it will, of course, also leave its mark on the culture of the samba school. For people interested in education in general, it will be important to trace the life histories of these efforts: How will they affect the intellectual development of their school-age participants? Will we see reversals of Piagetian stages? Will they develop pressures to withdraw from traditional schools? How will local schools try to adapt to the new pressure on them? But as an educational utopian I want something else. I want to know what kind of computer culture can grow in communities where there is not already a rich technophilic soil. I want to know and I want to help make it happen.

Let me say once more, the potential obstacle is not economic and it is not that computers are not going to be objects in people's everyday lives. They eventually will. They are already entering most workplaces and will eventually go into most homes just as TV sets now do, and in many cases initially for the same reasons. The obstacle to the growth of popular computer cultures is cultural, for example, the mismatch between the computer culture embedded in the machines of today and the cultures of the homes they will go into. And if the problem is cultural the remedy must be cultural.

The research challenge is clear. We need to advance the art of meshing computers with cultures so that they can serve to unite, hopefully without homogenizing, the fragmented subcultures that coexist counterproductively in contemporary society. For example, the gulf must be bridged between the technical-scientific and humanistic cultures. And I think that the key to constructing this bridge will be learning how to recast powerful ideas in computational form, ideas that are as important to the poet as to the engineer.

In my vision the computer acts as a transitional object to mediate relationships that are ultimately between person and person. There are mathophobes with a fine sense of moving their bodies, and there are mathophiles who have forgotten the sensory motor roots of their mathematical knowledge. The Turtle establishes a bridge. It serves as a common medium in which can be recast the shared elements of body geometry and formal geometry. Recasting juggling as structured programming can build a bridge between those who have a fine mathetic sense of physical skills and those who know how to go about organizing the task of writing an essay on history.

Juggling and writing an essay seem to have little in common if one looks at the *product*. But the processes of learning both skills have much in common. By creating an intellectual environment in which the emphasis is on process we give people with different skills and interests something to talk about. By developing expressive languages for talking about process and by recasting old knowledge in these new languages we can hope to make transparent the barriers separating disciplines. In the schools math is math and history is history and juggling is outside the intellectual pale. Time will tell whether schools can adapt themselves. What is more important is understanding the recasting of knowledge into new forms.

In this book we have seen complex interactions between new technologies and the recasting of the subject matters. When we discussed the use of the computer to facilitate learning Newton's laws of motion, we did not attempt to "computerize" the equations as they are found in a classical textbook. We developed a new conceptual framework for thinking about motion. For example, the concept of Turtle enabled us to formulate a qualitative component of Newtonian physics. The resulting reconceptualizing would be valid without a computer; its relation to the computer is not at all reductionist. But it is able to take advantage of the computer in ways in which other conceptualizations of physics could not, and thus gain in mathetic power. Thus, the whole process involves a dialectical interaction between new technologies and new ways of doing physics. The logic of these interactions is seen very clearly by looking at another item from my collection of good models for thinking about education.

Twenty years ago, parallel skiing was thought to be a skill attainable only after many years of training and practice. Today, it is routinely achieved during the course of a single skiing season. Some of the factors that contributed to this change are of a kind that fit into the traditional paradigms for educational innovation.

For example, many ski schools use a new pedagogical technique (the graduated length method—GLM) in which one first learns to ski using short skis and then gradually progresses to longer ones. But something more fundamental happened. In a certain sense what new skiers learn today so easily is not the same thing that their parents found so hard. All the goals of the parents are achieved by the children: The skiers move swiftly over the mountain with their skis parallel, avoiding obstacles and negotiating slalom gates. But the movements they make in order to produce these results are quite different.

When the parents learned to ski, both vacation skiers and Olympic champions used turning techniques based on a preparatory counterrotation, thought to be necessary for parallel turns. The realization that more direct movements could produce a more effective turn was a fundamental discovery, and it rapidly transformed skiing, both for the vacation skier and the champion. For the novice the new techniques meant more rapid learning, for the champion it meant more efficient movements, for the fashionable skier it meant more opportunities for elegant movements. Thus, at the heart of the change is a reconceptualization of skiing itself, not a mere change in pedagogy or technology. But in order to have a complete picture, we must also recognize a dialectical interaction between the content, the pedagogy, and the technology. For as ski movements were changing, skis and boots were changing too. New plastics allowed boots to become lighter and more rigid, and skis could be made more or less flexible. The direction of these changes was so synergistic with the new ski techniques that many ski instructors and ski writers attributed the change in skiing to the technology. Similarly, the use of short skis for instruction happened to be so highly adaptable to the new technology that many people sum up the ski revolution as the "move to GLM."

I like to think about the "ski revolution" because it helps me to think about the very complex junction we are at in the history of the "computer revolution." Today we hear a lot of talk about how "computers are coming" and a lot of talk about how they will change education. Most of the talk falls into two categories, one apparently "revolutionary" and the other "reformist." For many

185

revolutionaries, the presence of the computer will in itself produce momentous change: Teaching machines in the homes and computer networks will make school (as we know it) obsolete; reconceptualizations of physics are the furthest things from their minds. For the reformists, the computer will not abolish schools but will serve them. The computer is seen as an engine that can be harnessed to existing structures in order to solve, in local and incremental measures, the problems that face schools as they exist today. The reformist is no more inclined than the revolutionary to think in terms of reconceptualizing subject domains.

Our philosophy, both implicit and explicit, tries to avoid the two common traps: commitment to technological inevitability and commitment to strategies of incremental change. The technology itself will not draw us forward in any direction I can believe in either educationally or socially. The price of the education community's reactive posture will be educational mediocrity and social rigidity. And experimenting with incremental changes will not even put us in a position to understand where the technology is leading.

My own philosophy is revolutionary rather than reformist in its concept of change. But the revolution I envision is of ideas, not of technology. It consists of new understandings of specific subject domains and in new understandings of the process of learning itself. It consists of a new and much more ambitious setting of the sights of educational aspiration.

I am talking about a revolution in ideas that is no more reducible to technologies than physics and molecular biology are reducible to the technological tools used in the laboratories or poetry to the printing press. In my vision, technology has two roles. One is heuristic: The computer presence has catalyzed the emergence of ideas. The other is instrumental: The computer will carry ideas into a world larger than the research centers where they have incubated up to now.

I have suggested that the absence of a suitable technology has been a principle cause of the past stagnation of thinking about education. The emergence first of large computers and now of the microcomputer has removed this cause of stagnation. But there is another, secondary cause that grew like algae on a stagnant pond. We have to consider whether it will disappear with the condition that allowed its growth, or whether, like QWERTY, it will remain to strangle progress. In order to define this obstacle and place it in perspective, we shall pick out one of the salient ideas presented in earlier chapters and consider what besides technology is needed to implement it.

Out of the crucible of computational concepts and metaphors, of predicted widespread computer power and of actual experiments with children, the idea of Piagetian learning has emerged as an important organizing principle. Translated into practical terms this idea sets a research agenda concerned with creating conditions for children to explore "naturally" domains of knowledge that have previously required didactic teaching; that is, arranging for the children to be in contact with the "material"—physical or abstract—they can use for Piagetian learning. The prevalence of paired things in our society is an example of "naturally" occurring Piagetian material. The Turtle environments gave us examples of "artificial" (that is, deliberately invented) Piagetian material. Pairings and Turtles both owe their mathetic power to two attributes: Children relate to them, and they in turn relate to important intellectual structures. Thus pairing and Turtles act as transitional objects. The child is drawn into playing with pairs and with the process of pairing and in this play pairing acts as a carrier of powerful ideas—or of the germs from which powerful ideas will grow in the matrix of the child's active mind.

The attributes the Turtle shares with pairing might seem simple, but their realization draws upon a complex set of ideas, of kinds of expertise, and of sensitivities that can be broken down, though somewhat artificially, into three categories: knowledge about computers, knowledge about subject domains, and knowledge about people. The people knowledge I see as necessary to the design of good Piagetian material is itself complex. It includes the kinds of knowledge that are associated with academic psychology in all its branches—cognitive, personality, clinical, and so on—and also the more empathetic kinds possessed by creative artists and by people who "get along with children." In articulating these prerequisites for the creation of Piagetian material, we come face to face with

187

what I see as the essential remaining problem in regard to the future of computers and education: the problem of the supply of people who will develop these prerequisites.

This problem goes deeper than a mere short supply of such people. The fact that in the past there was no role for such people has been cast into social and institutional concrete; now there is a role but there is no place for them. In current professional definitions physicists think about how to do physics, educators think about how to teach it. There is no recognized place for people whose research is really physics, but physics oriented in directions that will be educationally meaningful. Such people are not particularly welcome in a physics department; their education goals serve to trivalize their work in the eyes of other physicists. Nor are they welcome in the education school—there, their highly technical language is not understood and their research criteria are out of step. In the world of education a new theorem for a Turtle microworld, for example, would be judged by whether it produced a "measurable" improvement in a particular physics course. Our hypothetical physicists will see their work very differently, as a theoretical contribution to physics that in the long run will make knowledge of the physical universe more accessible, but which in the short run would not be expected to improve performance of students in a physics course. Perhaps, on the contrary, it would even harm the student if injected as a local change into an educational process based on a different theoretical approach.

This point about what kind of discourse is welcome in schools of education and in physics departments is true more generally also. Funding agencies as well as universities do not offer a place for any research too deeply involved with the ideas of science for it to fall under the heading of education and too deeply engaged in an educational perspective for it to fall under the heading of science. It seems to be nobody's business to think in a fundamental way about science in relation to the way people think and learn it. Although lip service has been paid to the importance of science and society, the underlying methodology is like that of traditional education: one of delivering elements of ready-made science to a special audience. The concept of a *serious* enterprise of *making* science for the people is quite alien.

The computer by itself cannot change the existing institutional assumptions that separate scientist from educator, technologist from humanist. Nor can it change assumptions about whether science for the people is a matter of packaging and delivery or a proper area of serious research. To do any of these things will require deliberate action of a kind that could, in principle, have happened in the past, before computers existed. But it did not happen. The computer has raised the stakes both for our inaction and our action. For those who would like to see change, the price of inaction will be to see the least desirable features of the status quo exaggerated and even more firmly entrenched. On the other hand, the fact that we will be in a period of rapid evolution will produce footholds for institutional changes that might have been impossible in a more stable period.

The emergence of motion pictures as a new art form went hand in hand with the emergence of a new subculture, a new set of professions made up of people whose skills, sensitivities, and philosophies of life were unlike anything that had existed before. The story of the evolution of the world of movies is inseparable from the story of the evolution of the communities of people. Similarly, a new world of personal computing is about to come into being, and its history will be inseparable from the story of the people who will make it.

Epilogue:

The Mathematical Unconscious

REPRINTED HERE as an epilogue is my first discussion, written a few years ago, of an idea that developed into a central theme of this book: My rejection of the dichotomy opposing a stereotypically "disembodied" mathematics to activities engaging a full range of human sensitivities.* In the book I discuss this theme in the context of Turtle geometry. In the following pages the reader will find this theme embedded in reflections on the sources of mathematical pleasure.

It is deeply embedded in our culture that the appreciation of mathematical beauty and the experience of mathematical pleasure are accessible only to a minority, perhaps a very small minority, of the human race. This belief is given the status of a theoretical principle by Henri Poincaré, who has to be respected not only as one of the seminal mathematical thinkers of the century but also as one of the most thoughtful writers on the epistemology of the mathematical sciences. Poincaré differs sharply from prevalent trends in cognitive and educational psychology in his view of what makes a mathematician. For Poincaré the distinguishing feature of the mathematical mind is not logical but aesthetic. He also believes,

*I would like to thank the editors of the MIT press for their permission to reprint this essay which originally appeared as "Poincaré and the Mathematical Unconscious" in Judith Wechsler, ed., *Aesthetics In Science* (Cambridge, Mass.: MIT Press, 1978). I also want to thank Judith Wechsler for encouraging me to write this essay (which began as a guest lecture in one of her classes at MIT) and for much else as well.

but this is a separate issue, that this aesthetic sense is innate: Some people happen to be born with the faculty of developing an appreciation of mathematical beauty, and those are the ones who can become creative mathematicians. The others cannot.

This essay uses Poincaré's theory of mathematical creativity as an organizing center for reflections on the relationship between the logical and the extralogical in mathematics and on the relationship between the mathematical and the nonmathematical in the spectrum of human activities. The popular and the sophisticated wings of our culture almost unanimously draw these dichotomies in hard-edged lines. Poincaré's position is doubly interesting because in some ways he softens, and in some ways sharpens, these lines. They are softened when he attributes to the aesthetic an important functional role in mathematics. But the act of postulating a specifically mathematical aesthetic, and particularly an innate one, sharpens the separation between the mathematical and the nonmathematical. Is the mathematical aesthetic really different? Does it have common roots with other components of our aesthetic system? Does mathematical pleasure draw on its own pleasure principles or does it derive from those that animate other phases of human life? Does mathematical intuition differ from common sense in nature and form or only in content?

These questions are deep, complex, and ancient. My daring to address them in the space of a short essay is justified only because of certain simplifications. The first of these is a transformation of the questions, similar in spirit to Jean Piaget's way of transforming philosophical questions into psychogenetic ones to which experimental investigations into how children think become refreshingly relevant. By so doing, he has frequently enraged or bewildered philosophers, but has enriched beyond measure the scientific study of mind. My transformation turns Poincaré's theory of the highest mathematical creativity into a more mundane but more manageable theory of ordinary mathematical (and possibly nonmathematical) thinking.

Bringing his theory down to earth in this way possibly runs the risk of abandoning what Poincaré himself might have considered to be most important. But it makes the theory more immediately relevant, perhaps even quite urgent, for psychologists, educators, and others. For example, if Poincaré's model turned out to contain elements of a true account of ordinary mathematical thinking, it could follow that mathematical education as practiced today is totally misguided and even self-defeating. If mathematical aesthetics gets any attention in the schools, it is as an epiphenomenon, an icing on the mathematical cake, rather than as the driving force which makes mathematical thinking function. Certainly the widely practiced theories of the psychology of mathematical development (such as Piaget's) totally ignore the aesthetic, or even the intuitive, and concentrate on structural analysis of the logical facet of mathematical thought.

The destructive consequences of contemporary mathematics teaching can also be seen as a minor paradox for Poincaré. The fact that schools, and our culture generally, are so far from being nurturant of nascent mathematical aesthetic sense in children causes Poincaré's major thesis about the importance of aesthetics to undermine the grounds for believing in his minor thesis, which asserts the innateness of such sensibilities. If Poincaré is right about aesthetics, it becomes only too easy to see how the apparent rareness of mathematical talent could be explained without appeal to innateness.

These remarks are enough to suggest that the mundane transformation of Poincaré's theory might be a rich prize for educators even if it lost all touch with the processes at work in big mathematics. But perhaps we can have the best of both worlds. By adopting, as we shall, a more experiential mode of discussion through which theories about mathematical thinking can be immediately confronted with the reader's own mental processes, we do not, of course, renounce the possibility that the mathematical elite share similar experiences. On the contrary, that part of Poincaré's thinking which will emerge as most clearly valid in the ordinary context resonates strongly with modern trends which, in my view, constitute a paradigm shift in thinking about the foundations of mathematics. The concluding paragraphs of my essay will illustrate this resonance in the case of the Bourbaki theory of the structure of mathematics.

My goal here is not to advance a thesis with crisp formulations and rigorous argument, and it is certainly not to pass judgment on the correctness of Poincaré's theory. I shall be content (this is my second major simplification) to suggest to nonmathematical readers perceptions of, and a discourse about, mathematics which will place it closer than is commonly done to other experiences they know and enjoy. The major obstacle to doing so is a projection of mathematics which greatly exaggerates its logical face, much as the Mercator projection of the globe exaggerates the polar regions so that on the map northern Greenland becomes more imposing than equatorial Brazil. Thus our discussion will be aimed at distinguishing and relating what I shall call the extralogical face of mathematics and its logical face. I shall ignore distinctions which ought to be made within these categories. Mathematical beauty, mathematical pleasure, and even mathematical intuition will be treated almost interchangeably insofar as they are representatives of the extralogical. And, on the other side, we shall not separate such very different facets of the logical as the formalists' emphasis on the deductive process, Bertrand Russell's reductionist position (against which Poincaré fought so savagely), and Alfred Tarski's set theoretic semantics. These logical theories can be thrown together insofar as they have in common an intrinsic, autonomous view of mathematics. They deal with mathematics as self-contained, as justifying itself by formally defined (that is, mathematical) criteria of validity, and they ignore all reference of mathematics to anything outside itself. They certainly ignore phenomena of beauty and pleasure.

There is no theoretical tension in the fact that mathematical logicians ignore, as long as they do not deny, the extralogical. No one will call into question either the reality of the logical face of mathematics or the reality of mathematical beauty or pleasure. What Poincaré challenges is the possibility of understanding mathematical activity, the work of the mathematician solely, or even primarily, in logical terms without reference to the aesthetic. Thus his challenge is in the field of psychology, or the theory of mind, and, as such, has wider reverberations than the seemingly specialized problem of understanding mathematical thinking: His challenge calls into question the separation within psychology of cognitive functions, defined by their opposition to considerations of affect, of feeling, of sense of beauty.

I shall, on the whole, side with Poincaré against the possibility of a "purely cognitive" theory of mathematical thinking but express reservations about the high degree of specificity he attributes to the mathematical. But first I must introduce another of the themes of Poincaré's theory. This is the role and the nature of the unconscious.

As the aesthetic versus the logical leads us to confront Poincaré with cognitive psychology, so the unconscious versus the conscious leads to a confrontation with Freud. Poincaré is close to Freud in clearly postulating two minds (the conscious and the unconscious) each governed by its own dynamic laws, each able to carry out different functions with severely limited access to the other's activities. As we shall see, Poincaré is greatly impressed by the way in which the solution to a problem on which one has been working at an earlier time often comes into consciousness unannounced, and almost ready-made, as if produced by a hidden part of the mind. But Poincaré's unconscious is very different from Freud's. Far from being the site of prelogical, sexually charged, primary processes, it is rather like an emotionally neutral, supremely logical, combinatoric machine.

The confrontation of these images of the unconscious brings us back to our questions about the nature of mathematics itself. The logical view of mathematics is definitionally discorporate, detached from the body and molded only by an internal logic of purity and truth. Such a view would be concordant with Poincaré's neutral unconscious rather than with Freud's highly charged, instinct-ridden dynamics. But Poincaré himself, as I have already remarked, rejects this view of mathematics; even if it could be maintained (which is already dubious) as an image of the finished mathematical product, it is totally inadequate as an account of the productive process through which mathematical truths and structures emerge. In its most naive form the logical image of mathematics is a deductive system in which new truths are derived from previously derived truths by means of rigorously reliable rules of inference. Although less naive logicist theses cannot be demolished quite so easily, it is relevant to notice the different ways in which this account of mathematics can be criticized. It is certainly incomplete since it fails to explain the process of choice determining how deductions are made and which of those made are pursued. It is misleading in that the rules of inference actually used by mathematicians would, if applied incautiously, quickly lead to contradictions and paradoxes. Finally, it is factually false as a description in that it provides no place for the as yet undebugged partial results with which the actual mathematician spends the most time. Mathematical work does not proceed along the narrow logical path of truth to truth to truth, but bravely or gropingly follows deviations through the surrounding marshland of propositions which are neither simply and wholly true nor simply and wholly false.

Workers in artificial intelligence have patched up the first of these areas of weakness, for example, by formalizing the process of setting and managing new problems as part of the work of solving a given one. But if the new problems and the rules for generating them are cast in logical terms, we see this as, at best, the replacement of a static logic by a dynamic one. It does not replace logic by something different. The question at issue here is whether even in the course of working on the most purely logical problem the mathematician evokes processes and sets problems which are not themselves purely logical.

The metaphor of wandering off the path of truth into surrounding marshlands has the merit, despite its looseness, of sharply stating a fundamental problem and preoccupation of Poincaré's: the problem of guidance, or one might say, of "navigation in intellectual space." If we are content to churn out logical consequences, we would at least have the security of a safe process. In reality, according to Poincaré, the mathematician is guided by an aesthetic sense: In doing a job, the mathematician frequently has to work with propositions which are false to various degrees but does not have to consider any that offend a personal sense of mathematical beauty.

Poincaré's theory of how the aesthetic guides mathematical work divides the work into three stages. The first is a stage of deliberate, conscious analysis. If the problem is difficult, the first stage will

195

never, according to Poincaré, yield the solution. Its role is to create the elements out of which the solution will be constructed. A stage of unconscious work, which might appear to the mathmatician as temporarily abandoning the task or leaving the problem to incubate, has to intervene. Poincaré postulates a mechanism for the incubation. The phenomenological view of abandonment is totally false. On the contrary, the problem has been turned over to a very active unconscious which relentlessly begins to combine the elements supplied to it by the first, conscious state of the work. The unconscious mind is not assumed to have any remarkable powers except concentration, systematic operation, and imperviousness to boredom, distractions, or changes of goal. The product of the unconscious work is delivered back to the conscious mind at a moment which has no relation to what the latter is doing. This time the phenomenological view is even more misleading since the finished piece of work might appear in consciousness at the most surprising times, in apparent relation to quite fortuitous events.

How does the unconscious mind know what to pass back to the conscious mind? It is here where Poincaré sees the role of the aesthetic. He believes, as a matter of empirical observation, that ideas passed back are not necessarily correct solutions to the original problem. So he concludes that the unconscious is not able to rigorously determine whether an idea is correct. But the ideas passed up do always have the stamp of mathematical beauty. The function of the third stage of the work is to consciously and rigorously examine the results obtained from the unconscious. They might be accepted, modified, or rejected. In the last case the unconscious might once more be called into action. We observe that the model postulates a third agent in addition to the conscious and unconscious minds. This agent is somewhat akin to a Freudian censor; its job is to scan the changing kaleidoscope of unconscious patterns allowing only those which satisfy its aesthetic criteria to pass through the portal between the minds.

Poincaré is describing the highest level of mathematical creativity, and one cannot assume that more elementary mathematical work follows the same dynamic processes. But in our own striving toward a theory of mathematical thinking we should not assume the contrary either, and so it is encouraging to see even very limited structural resemblances between the process as described by Poincaré and patterns displayed by nonmathematicians whom we asked to work on mathematical problems in what has come, at MIT, to be called "Loud Thinking," a collection of techniques designed to elicit productive thought (often in domains, such as mathematics, they would normally avoid) and make as much of it as possible explicit. The example that follows illustrates aspects of what the very simplest kind of aesthetic guidance of thought might be. The subjects in the experiment clearly proceed by a combinatoric, such as that which Poincaré postulates in his second stage, until a result is obtained which is satisfactory on grounds that have at least as much claim to be called aesthetic as logical. The process does differ from Poincaré's description in that it remains on the conscious level. This could be reconciled with Poincaré's theory in many ways: One might argue that the number of combinatorial actions needed to generate the acceptable result is too small to require passing the problem to the unconscious level, or that these nonmathematicians lack the ability to do such work unconsciously. In any case, the point of the example (indeed, of this essay as a whole) is not to defend Poincaré in detail but to illustrate the concept of aesthetic guidance.

The problem on which the subjects were asked to work was the proof that the square root of 2 is irrational. The choice is particularly appropriate here because this theorem was selected by the English mathematician G. H. Hardy as a prime example of mathematical beauty, and consequently it is interesting, in the context of a nonelitist discussion of mathematical aesthetics, to discover that many people with very little mathematical knowledge are able to discover the proof if emotionally supportive working conditions encourage them to keep going despite mathematical reticence. The following paragraphs describe an episode through which almost all the subjects in our investigation passed. To project ourselves into this episode, let us suppose that we have set up the equation:

$$\sqrt{2} = p/q \quad \text{where } p \text{ and } q \text{ are whole numbers}$$

Let us also suppose that we do not really believe that $\sqrt{2}$ can be so expressed. To prove this, we seek to reveal something bizarre, in fact contradictory, behind the impenetrably innocent surface impression of the equation. We clearly have to do with an interplay of latent and manifest contents. What steps help in such cases?

Almost as if they had read Freud, many subjects engage in a process of mathematical "free association," trying in turn various transformations associated with equations of this sort. Those who are more sophisticated mathematically need a smaller number of tries, but none of the subjects seem to be guided by a prevision of where the work will go. Here are some examples of transformations in the order they were produced by one subject:

$$\sqrt{2} = p/q$$
$$\sqrt{2} \times q = p$$
$$p = \sqrt{2} \times q$$
$$(\sqrt{2})^2 = (p/q)^2$$
$$2 = p^2/q^2$$
$$p^2 = 2q^2$$

All subjects who have become more than very superficially involved in the problem show unmistakable signs of excitement and pleasure when they hit on the last equation. This pleasure is not dependent on knowing (at least consciously) where the process is leading. It happens before the subjects are able to say what they will do next, and, in fact, it happens even in cases where no further progress is made at all. And the reaction to $p^2 = 2q^2$ is not merely affective; once this has been seen, the subjects scarcely ever look back at any of the earlier transformations or even at the original equations. Thus there is something very special about $p^2 = 2q^2$. What is it? We first concentrate on the fact that it undoubtedly has a pleasurable charge and speculate about the sources of the charge. What is the role of pleasure in mathematics?

Pleasure is, of course, often experienced in mathematical work, as if one were rewarding oneself when one achieves a desired goal after arduous struggle. But it is highly implausible that this actual equation was anticipated here as a preset goal. If the pleasure was that of goal achievement, the goal was of a very different, less formal, I would say "more aesthetic" nature than the achievement of a particular equation. To know exactly what it is would require much more knowledge about the individual subjects than we can include here. It is certainly different from subject to subject and even multiply overdetermined in each subject. Some subjects explicitly set themselves the goal: "Get rid of the square root." Other subjects did not seem explicitly to set themselves this goal but were nevertheless pleased to see the square root sign go away. Others, again, made no special reaction to the appearance of $2 = p^2/q^2$ until this turned into $p^2 = 2q^2$. My suggestion is that the elimination of the root sign for the obvious, simple, instrumental purpose is only part of a more complex story: The event is resonant with several processes which might or might not be accessible to the conscious mind and might or might not be explicitly formulated as goals. I suggest, too, that some of these processes tap into other sources of pleasure, more specific and perhaps even more primitive than the generalized one of goal attainment. To make these suggestions more concrete, I shall give two examples of such pleasure-giving processes.

The first example is best described in terms of the case frame type of calculus of situations characterizing recent thinking in artificial intelligence. The original equation is formalized as a situation frame with case slots for "three actors," of which the principle or "subject" actor is $\sqrt{2}$. The other two actors, p and q, are subordinate dummy actors whose roles are merely to make assertions about the subject actor. When we turn the situation into $p^2 = 2q^2$, it is as sharply different as in a figure/ground reversal or the replacement of a screen by a face in an infant's perception of peek-a-boo. Now p has become the subject, and the previous subject, $\sqrt{2}$, has vanished. Does this draw on the pleasure sources that make infants universally enjoy peek-a-boo?

The other example of what might be pleasing in this process comes from the observation that 2 has not vanished away completely without trace. The 2 is still visible in $p^2 = 2q^2$! However, these two occurences of 2 are so very different in role that identifying them gives the situation a quality of punning, or "condensation" at least somewhat like that which Freud sees as fundamental to the effectiveness of wit. The attractiveness and plausibility of this suggestion comes from the possibility of seeing condensation in very many mathematical situations. Indeed, the very central idea of abstract mathematics could be seen as condensation: The "abstract" description simultaneously signifies very different "concrete" things. Does this allow us to conjecture that mathematics shares more with jokes, dreams, and hysteria than is commonly recognized?

It is of course dangerous to go too far in the direction of presenting the merits of $p^2 = 2q^2$ in isolation from its role in achieving the original purpose, which was not to titillate the mathematical pleasure senses but to prove that 2 is irrational. The statement of the previous two paragraphs needs to be melded with an understanding of how the work comes to focus on $p^2 = 2q^2$ through a process not totally independent of recognizing it as a subgoal of the supergoal of proving the theorem. How do we integrate the functional with the aesthetic? The simplest gesture in this direction for those who see the eminently functional subgoal system as the prime mover is to enlarge the universe of discourse in which subgoals can be formulated. Promoting a subordinate character (that is, p) on the problem scene to a principal role is, within an appropriate system of situation frames, as well-defined a subgoal as, say, finding the numerical solution of an equation. But we are now talking about goals which have lost their mathematical specificity and may be shared with nonmathematical situations of life or literature. Taken to its extreme, this line of thinking leads us to see mathematics, even in its detail, as an acting out of something else: The actors may be mathematical objects, but the plot is spelled out in other terms. Even in its less extreme forms this shows how the aesthetic and the functional can enter into a symbiotic relationship of, so to speak, mutual exploitation. The mathematically functional goal is achieved through a play of subgoals formulated in another, nonmathematical discourse, drawing on corresponding extramathematical knowledge. Thus the functional exploits the aesthetic. But to the extent we see (here in a very Freudian spirit) the mathematical process itself as acting out premathematical processes, the reverse is also true.

These speculations go some (very little) way toward showing how Poincaré's mathematical aesthetic sentinel could be reconciled with existing models of thinking to the enrichment of both. But the attempt to do so very sharply poses one fundamental question about the relationship between the functional and the aesthetic and hedonistic facets not only of mathematics but of all intellectual work. What is it about each of these that makes it able to serve the other? Is it not very strange that knowledge, or principles of appreciation, useful outside of mathematics, should have application within? The answer must lie in a genetic theory of mathematics. If we adopt a Platonic (or logical) view of mathematics as existing independently of any properties of the human mind, or of human activity, we are forced to see such interpretations as highly unlikely. In the remaining pages I shall touch on a few more examples of how mathematics can be seen from a perspective which makes its relationship to other human structures more natural. We begin by looking at another episode of the story about the square root of 2.

Our discussion of $p^2 = 2q^2$ was almost brutally nonteleological in that we discussed it from only one side, the side from which it came, pretending ignorance of where it was going. We now remedy this by seeing how it serves the original intention of the work, which was to find a contradiction in the assumption $\sqrt{2} = p/q$. It happens that there are several paths one can take to this goal. Of these I shall contrast two which differ along a dimension one might call "gestalt versus atomistic" or "aha-single-flash-insight versus step-by-step reasoning." The step-by-step form is the more classical (it is attributed to Euclid himself) and proceeds in the following manner. We can read off from $p^2 = 2q^2$ that p^2 is even. It follows that p is even. By definition this means that p is twice some other whole number which we can call r. So:

$$p = 2r$$
$$p^2 = 4r^2$$
$$2q^2 = 4r^2 \quad \text{remember: } p^2 = 2q^2 (!)$$
$$q^2 = 2r^2$$

and we deduce that q is also even. But this at last really is manifestly bizarre since we chose p and q in the first place and could, had we wished, have made sure that they had no common factor. So there is a contradiction.

Before commenting on the aesthetics of this process, we look at the "flash" version of the proof. It depends on having a certain perception of whole numbers, namely, as unique collections of prime factors: $6 = 3 \times 2$ and $36 = 3 \times 3 \times 2 \times 2$. If you solidly possess this frame for perceiving numbers, you probably have a sense of immediate perception of a perfect square (36 or p^2 or q^2) as an even set. If you do not possess it, we might have to use step-by-step arguments (such as let $p = p_1 p_2 \ldots p_k$, so that $p^2 = p_1 p_1 p_2 p_2 \ldots p_k p_k$), and this proof then becomes even more atomistic and certainly less pleasing than the classical form. But if you do see (or train yourself to see) p^2 and q^2 as even sets, you will also see $p^2 = 2q^2$ as making the absurd assertion that an even set (p^2) is equal to an odd set (q^2 and one additional factor: 2). Thus given the right frames for perceiving numbers, $p^2 = 2q^2$ is (or so it appears phenomenologically) directly perceived as absurd.

Although there is much to say about the comparative aesthetics of these two little proofs, I shall concentrate on just one facet of beauty and pleasure found by some subjects in our experiments. Many people are impressed by the brilliance of the second proof. But if this latter attracts by its cleverness and immediacy, it does not at all follow that the first loses by being (as I see it) essentially serial. On the contrary, there is something very powerful in the way one is captured and carried inexorably through the serial process. I do not merely mean that the proof is rhetorically compelling when presented well by another person, although this is an important factor in the spectator sport aspect of mathematics. I mean rather that you need very little mathematical knowledge for the steps to be forced moves, so that once you start on the track you will find that you generate the whole proof.

One can experience the process of inevitability in very different ways with very different kinds of affect. One can experience it as being taken over in a relationship of temporary submission. One can experience this as surrender to mathematics, or to another person, or of one part of oneself to another. One can experience it not as submission but as the exercise of an exhilarating power. Any of these can be experienced as beautiful, as ugly, as pleasurable, as repulsive, or as frightening.

These remarks, although they remain at the surface of the phenomenon, suffice to cast serious doubt on Poincaré's reasons for believing that the faculty for mathematical aesthetic is inborn and independent of other components of the mind. They suggest too many ways in which factors of a kind Poincaré does not consider might, in principle, powerfully influence whether an individual finds mathematics beautiful or ugly and which kinds of mathematics he will particularly relish or revile. To see these factors a little more clearly, let us leave mathematics briefly to look at an example from a very sensitive work of fiction: Robert Pirsig's *Zen and the Art of Motorcycle Maintenance.* The book is a philosophical novel about different styles of thought. The principal character, who narrates the events, and his friend John Sutherland are on a motorcycling vacation which begins by riding from the east coast to Montana. Some time before the trip recounted in the book, John Sutherland had mentioned that his handlebars were slipping. The narrator soon decided that some shimming was necessary and proposed cutting shim stock from an aluminum beer can. "I thought this was pretty clever myself," he says, describing his surprise at Sutherland's reaction which brought the friendship close to rupture. To Sutherland the idea was far from clever; it was unspeakably offensive. The narrator explains: "I had had the nerve to propose repair of his new eighteen-hundred-dollar BMW, the pride of a half-century of German mechanical finesse, with a piece of old beer can!" But for the narrator there is no conflict; on the contrary: "Beer can aluminum is soft and sticky as metals go. Perfect for the application . . . in other words any true German mechanic with half a century of mechanical finesse behind him, would have concluded that this particular solution to this particular technical problem was perfect." The difference proves to be unbridgeable and emotionally explosive. The friendship is saved only by a tacit agreement never again to discuss maintenance and repair of the motorcycles even though the two friends are close enough to one another and to their motorcycles to embark together on the long trip described in the book.

Sutherland's reaction would be without consequence for our problem if it showed stupidity, ignorance, or an idiosyncratic quirk about ad hoc solutions to repair problems. But it goes deeper than any of these. Pirsig's accomplishment is to show us the coherence in many such incidents. This accomplishment is quite impressive. Pirsig presents us with materials so rich that we can use them to appreciate kinds of coherence implicit in the incidents which are rather different from the one advanced by Pirsig himself. Here I want to touch briefly on two analogies between the story of Sutherland and the shim stock and issues we have discussed about mathematics: first, the relationship between aesthetics and logic in thinking about mathematics as well as motorcycles, and second, the lines of continuity and discontinuity between mathematics or motorcycles and everything else.

It is clear from the shim stock incident itself, and much more so from the rest of the book, that the continuity between man, machine, and natural environment is very different for each of Pirsig's characters and that these differences deeply affect their aesthetic appreciation. For the narrator, the motorcycle is continuous with the world not only of beer cans but more generally the world of metals (taken as substance). In this world, the metal's identity is not reducible to a particular embodiment of the metal in a motorcycle or in a beer can. Nor can any identity be reduced to a particular instance of it. For Sutherland, on the contrary, this continuity is not merely invisible, but he has a strong investment in maintaining the boundaries between what the narrator sees as superficial manifestations of the same substance.

For Sutherland, the motorcycle is not only in a world apart from beer cans; it is even in a world apart from other machines, a fact that enables him to relate without conflict to this piece of technology as a means to escape from technology. We could deepen the analysis of the investments of these two characters in their respective positions by noting their very different involvements in work and society. The narrator is part of industrial society (he works for a computer company) and is forced to seek his own identity (as he seeks the identity of metal) in a sense of his *substance* which lies beyond the particular form into which he has been molded. Like malleable metal, he is something beyond and perhaps better than the form which is now imposed on him. He certainly does not define himself as a writer of computer manuals. His friend Sutherland on the other hand is a musician and is much more able to take his work as that which structures his image of himself in the same way that he takes a motorcycle as a motorcycle and a beer can as a beer can.

We need not pursue these questions of essence and accident much further to make the important point, and a point which is widely ignored: If styles of involvement with motorcycle maintenance are so intricately interwoven with our psychological and social identities, one would scarcely expect this to be less true about the varieties of involvements of individuals with mathematics.

These ideas about the relationship of mathematical work with the whole person were illuminated earlier in this book by Turtle geometry, as it is used with the LOGO programming language. These experiments express a critique of traditional school mathematics (which applies no less to the so-called new math than to the old). A description of traditional school mathematics in terms of the concepts we have developed in this essay would reveal it to be a caricature of mathematics in its depersonalized, purely logical, "formal" incarnation. Although we can document progress in the rhetoric of math teachers (teachers of the new math are taught to speak in terms of "understanding" and "discovery"), the problem remains because of what they are teaching.*

In Turtle geometry we create an environment in which the child's task is not to learn a set of formal rules but to develop sufficient insight into the way he moves in space to allow the transposition of this self-knowledge into programs that will cause a Turtle to move. By now the reader of this book is very familiar with the potential of this cybernetic animal. But what I would like to do here is recall and underscore two closely related aspects of Turtle geometry which are directly relevant to the concerns of this essay. The first is the development of an ego-syntonic mathematics, indeed, of a "body-syntonic" mathematics; the second is the development of a

*The following paragraphs have been modified for continuity with this book.

context for mathematical work where the aesthetic dimension (even in its narrowest sense of "the pretty") is continually placed in the forefront.

We shall give a single example which illuminates both of these aspects: an example of a typical problem that arises when a child is learning Turtle geometry. The child has already learned how to command the Turtle to move forward in the direction that it is facing and to pivot around its axis, that is, to turn the number of degrees right or left that the child has commanded. With these commands the child has written programs which cause the Turtle to draw straight line figures. Sooner or later the child poses the question: "How can I make the Turtle draw a circle?" In LOGO we do not provide "answers," but encourage learners to use their own bodies to find a solution. The child begins to walk in circles and discovers how to make a circle by going forward a little and turning a little, by going forward a little and turning a little. Now the child knows how to make the Turtle draw a circle: Simply give the Turtle the same commands one would give oneself. Expressing "go forward a little, turn a little" comes out in Turtle language as RE-PEAT [FORWARD 1 RIGHT TURN 1]. Thus we see a process of geometrical reasoning that is both ego syntonic and body syntonic. And once the child knows how to place circles on the screen with the speed of light, an unlimited palette of shapes, forms, and motion has been opened. Thus the discovery of the circle (and, of course, the curve) is a turning point in the child's ability to achieve a direct aesthetic experience through mathematics.

In the above paragraph it sounds as though ego-syntonic mathematics was recently invented. This is certainly not the case and, indeed, would contradict the point made repeatedly in this essay that the mathematics of the mathematician is profoundly personal. It is also not the case that we have invented ego syntonic mathematics for children. We have merely given children a way to reappropriate what was always theirs. Most people feel that they have no "personal" involvement with mathematics, yet as children they constructed it for themselves. Jean Piaget's work on genetic epistemology teaches us that from the first days of life a child is engaged in an enterprise of extracting mathematical knowledge from the intersection of body with environment. The point is that, whether we intend it or not, the teaching of mathematics, as it is traditionally done in our schools, is a process by which we ask the child to forget the natural experience of mathematics in order to learn a new set of rules.

This same process of forgetting extralogical roots has until very recently dominated the formal history of mathematics in the academy. In the early part of the twentieth century, formal logic was seen as synonymous with the foundation of mathematics. Not until Bourbaki's structuralist theory appeared do we see an internal development in mathematics which opens mathematics up to "remembering" its genetic roots. This "remembering" was to put mathematics in the closest possible relationship to the development of research about how children construct their reality.

The consequences of these currents and those we encountered earlier in cognitive and dynamic psychology place the enterprise of understanding mathematics at the threshold of a new period heralded by Warren McCulloch's epigrammatic assertion that neither man nor mathematics can be fully grasped separately from the other. When asked what question would guide his scientific life, McCulloch answered: "What is a man so made that he can understand number and what is number so made that a man can understand it?"

Afterword and Acknowledgments

IN 1964 I moved from one world to another. For the previous five years I had lived in Alpine villages near Geneva, Switzerland, where I worked with Jean Piaget. The focus of my attention was on children, on the nature of thinking, and on how children become thinkers. I moved to MIT into an urban world of cybernetics and computers. My attention was still focused on the nature of thinking, but now my immediate concerns were with the problem of Artificial Intelligence: How to make machines that think?

Two worlds could hardly be more different. But I made the transition because I believed that my new world of machines could provide a perspective that might lead to solutions to problems that had eluded us in the old world of children. Looking back I see that the cross-fertilization has brought benefits in both directions. For several years now Marvin Minsky and I have been working on a general theory of intelligence (called "The Society Theory of Mind") which has emerged from a strategy of thinking simultaneously about how children do and how computers might think.

Minsky and I, of course, are not the only workers to have drawn on the theory of computation (or information processing) as a source of models to be used in explaining psychological phenomena. On the contrary, this approach has been taken by such people as Warren McCulloch, Allen Newell, Herbert Simon, Alan Turing, Norbert Wiener, and quite a number of younger people. But the point of departure of this book is a point of view—first articulated jointly with Minsky—that separates us quite sharply from most other members of this company: that is to say, seeing ideas from computer science not only as *instruments of explanation* of how learning and thinking in fact do work, but also as *instruments of change* that might alter, and possibly improve, the way people learn and think.

The book grew out of a project designed to explore this concept by giving children access to "the best of computer science" including some of its best technology and some of its best ideas. At the heart of the project was the creation of a children's learning environment in the same building that houses MIT's Artificial Intelligence Laboratory and Laboratory for Computer Science (Project MAC). We hoped that by bringing children and people interested primarily in children into this world of computers and computerists, we would create conditions for a flow of ideas into thinking about education.

I shall not try to describe all that happened in the course of this project or all that was learned from it, but I shall concentrate on some personal reflections. Readers who want to know more about the project itself will find pointers to other publications in the notes at the end of the book.

The project is really an experiment in cultural interaction. It set out to grow a new "education culture" in an environment permeated with a particular form of "computer culture." Too many people were involved for me even to know all their names. The interchanges of ideas took place much more in conversations in the quiet of after-midnight hours (for this is a computer culture that does not respect the conventional clock cycles) than in organized seminars or written papers. In early drafts I attempted to chronicle the growth of the culture. But it proved too difficult and in the end I wrote the book in a very personal style. This has a certain advantage in allowing me to give freer reign to my own personal interpretations of ideas and incidents that other participants might well see very differently. I hope that it does not obscure my sense of belonging to a communtity and of expressing a set of shared ideas. I regret that space does not permit me to show how some of these ideas have been picked up by others and elaborated into much more advanced forms.

Marvin Minsky was the most important person in my intellectual life during the growth of the ideas in this book. It was from him that I first learned that computation could be more than a theoretical science and a practical art: It can also be the material from which to fashion a powerful and personal vision of the world. I have since encountered several people who have done this successfully and in an inspirational way. Of these, one who stands out because he has so consistently turned his personal computational vision to thinking about children is Alan Kay. During the whole decade of the 1970s, Kay's research group at the Xerox Palo Alto Research Center and our group at MIT were the only American workers on computers for children who made a clear decision that significant research could not be based on the primitive computers that were then becoming available in schools, resource centers, and education research laboratories. For me, the phrase "computer as pencil" evokes the kind of uses I imagine children of the future making of computers. Pencils are used for scribbling as well as writing, doodling as well as drawing, for illicit notes as well as for official assignments. Kay and I have shared a vision in which the computer would be used as casually and as personally for an even greater diversity of purposes. But neither the school computer terminal of 1970 nor the Radio Shack home computer of 1980 have the power and flexibility to provide even an approximation of this vision. In order to do so, a computer must offer far better graphics and a far more flexible language than computers of the 1970s can provide at a price schools and individuals can afford.

In 1967, before the children's laboratory at MIT had been officially formed, I began thinking about designing a computer language that would be suitable for children. This did not mean that it should be a "toy" language. On the contrary, I wanted it to have the power of professional programming languages, but I also wanted it to have easy entry routes for nonmathematical beginners. Wallace Feurzeig, head of the Educational Technology Group at the research firm of Bolt Beranek and Newman, quickly recognized the merit of the idea and found funding for the first implementation and trial of the language. The name LOGO was chosen for the new language to suggest the fact that it is primarily symbolic and only secondarily quantitative. My original design of the language was greatly improved in the course of discussions with Daniel Bobrow, who had been one of the first graduate students in the MIT Artificial Intelligence group, Cynthia Solomon, and Richard Grant, all of whom were working at that time at Bolt Baranek and Newman. Most subsequent development of the LOGO language, which has gone through several rounds of "modernization," took place at MIT. Of the very many people who contributed to it I can list only a few: Harold Abelson, Bruce Edwards, Andrea diSessa, Gary Drescher, Ira Goldstein, Mark Gross, Ed Hardebeck, Danny Hillis, Bob Lawler, Ron Lebel, Henry Lieberman, Mark Miller, Margaret Minsky, Cynthia Solomon, Wade Williams, and Terry Winograd. For many years Ron Lebel was the chief systems programmer in charge of LOGO development. But the people who worked directly on LOGO form only the tip of an iceberg: The influence of the MIT community on LOGO went much deeper.

Our Artificial Intelligence Laboratory has always been near the center of a movement, strongly countercultural in the larger world of computers, that sees programming languages as heavily invested with epistemological and aesthetic commitments. For me this "Whorfian" view has been best articulated in the work of three computer scientists who were graduate students at the time LOGO was in formation: Carl Hewitt, Gerald Sussman, and Terry Winograd. But it goes back to the founders of the MIT Artificial Intelligence group, Marvin Minsky and John McCarthy, and owes much to the tradition of "hackers" of whom I feel most directly the influence of William Gosper and Richard Greenblatt. In the cultural atmosphere created by such people it was as unacceptable for children to enter the computer culture by learning computer languages such as BASIC as it would be to confine their access to English poetry to pidgin English translations.

I have always considered learning a hobby and have developed many insights into its nature by cultivating a sensitivity to how I go about doing it. Thus, I have perhaps engaged in deliberate learning of a wider range of material than most people. Examples of things I have learned in this spirit include chapters of science (like thermodynamics), reading Chinese characters, flying airplanes, cooking in various cuisines, performing circus arts such as juggling, and even two bouts of living for several weeks with distorting spectacles (on one occasion left-right reversing glasses, on the other a rather complex prismatic distortion of the visual field). Part of what I found so attractive about the Artificial Intelligence community was a shared interest in this approach to using one's self as a source of insight into psychological processes and a particular interest in observing oneself engaged in skilled activities. Here again I owe debts to many people and am able to single out only those whose contributions were most salient: Howard Austin, Jeanne Bamberger, Ira Goldstein, Bob Lawler, Gerald Sussman, and the graduate students who took part in my "loud thinking seminars" where such methods were explored. My approach to "loud thinking" acquired greater sophistication during a period of collaboration with Donald Schon and Benson Snyder and in interaction with a number of psychologists including Edith Ackermann, Daniel Bobrow, Howard Gruber, Annette Karmiloff-Smith, and Donald Norman.

All these influences entered into the emergence of a learning/teaching methodology in the computational environments we were building for children. The person closest to me in this work was Cynthia Solomon. As in the case of Marvin Minsky, my collaboration with her was so close over so long a period that I find it impossible to enumerate the substantial contributions she made. Solomon was also the first to develop an intellectually coherent methodology for training teachers to introduce children to computers and is still one of the few people to have approached this problem with the seriousness it deserves.

Many people contributed ideas about teaching children LOGO. Ira Goldstein undertook the difficult problem of developing a theoretical framework for the instructional process and was followed in this work by Mark Miller. Others approached teaching in a more pragmatic spirit. Special contributions have been made by Howard Austin, Paul Goldenberg, Gerianne Goldstein, Virginia Grammar, Andree Green, Ellen Hildreth, Kiyoko Okumura, Neil Rowe, and Dan Watt. Jeanne Bamberger developed methods for using LOGO in musical learning and in increasing teachers' sensitivity to their own thinking.

A central idea behind our learning environments was that children would be able to use powerful ideas from mathematics and science as instruments of personal power. For example, geometry would become a means to create visual effects on a television screen. But achieving this often meant developing new topics in mathematics and science, an enterprise that was possible only because we were working within an institution rich in creative mathematical talent. The task is of a new kind: It consists of doing what is really original research in mathematics or science but in directions chosen because they lead to more comprehensible or more learnable forms of knowledge and not for the kinds of reasons that typically motivate mathematical research. Many students and faculty members at MIT contributed to this work, but two stand out as professionals in the area: Harold Abelson, a mathematician, and Andrew diSessa, a physicist.

Many LOGO workers contributed to the aesthetic of the Turtle drawings. Those who most influenced me were Cynthia Solomon, Ellen Hildreth and Ilse Schenck (who arranged the garden and birds in this book).

In this book I write about children but, in fact, most of the ideas expressed are relevant to how people learn at any age. I make specific references to children as a reflection of my personal conviction that it is the very youngest who stand to gain the most from change in the conditions of learning. Most of the children who collaborated with us were of mid-elementary school age. Radia Perlman was the first to explore techniques for working with much younger children, as young as four years of age. Abelson and diSessa have specialized in work with older students of high school and college age. Gary Drescher, Paul Goldenberg, Sylvia Weir, and Jose Valente are among those who have pioneered teaching LOGO to severely handicapped children. Bob Lawler carried out the first, and so far the only, example of a different kind of learning experiment, a kind that I think will become very important in the future. In Lawler's study, a child was observed "full time" during a six-month period so as to capture not only the learning that took place in contrived situations but all the overt learning that took place during that period. I have also been influenced by another study on "natural learning" now being conducted as part of research by Lawrence Miller for his thesis at Harvard. Both Lawler and Miller provided data for a general intellectual position that underlies this book: The best learning takes place when the learner takes charge. Edwina Michner's Ph.D. thesis was a learning study of a very different sort, an attempt to characterize some of the mathematical knowledge that the mathematical culture does not write down in its books.

I have acknowledged intellectual obligations to many people. I have to thank most of them for something else as well: for support and for patience with my too often disorganized working style. I am deeply grateful to everyone who put up with me, especially Gregory Gargarian who had the very difficult jobs of maintaining the organization of the LOGO Laboratory and of entering and updating many successive versions of this book in the computer files. In addition to his competence and professionalism, his friendship and support have made easier many moments in the writing of this book.

MIT has provided a highly stimulating intellectual environment. Its administrative environment is also very special in allowing out-of-the-ordinary projects to flourish. Many people have helped in an administrative capacity: Jerome Wiesner, Walter Rosenblith, Michael Dertouzos, Ted Martin, Benson Snyder, Patrick Winston, Barbara Nelson, Eva Kampits, Jim McCarthy, Gordon Oro, Russel Noftsker, George Wallace, Elaine Medverd, and surely others. Of these I owe a very special debt to Eva Kampits, who was once my secretary and is now Dr. Kampits.

The LOGO project could not have happened without support of a different kind than I have mentioned until now. The National Science Foundation has supported the work on LOGO since its inception. I want also to mention some of the Foundation's individuals whose imaginative understanding made it possible for us to do our work: Dorothy Derringer, Andrew Molnar, and Milton Rose. The value of the support given by such people is moral as well as material, and I would include in this category Marjorie Martus at the Ford Foundation, Arthur Melmed at the National Institute of Education, Alan Ditman at the Bureau for the Education of the Handicapped, and Alfred Riccomi of Texas Instruments. I would also most especially include three individuals who have given us moral and material support: Ida Green, Erik Jonsson, and Cecil Green all from Dallas, Texas. It has been a particularly rich experience for me to work closely with Erik Jonsson on developing a project using computers in the Lamplighter School in Dallas. I have come to appreciate his clarity of thought and breadth of vision and to think of him as a colleague and a friend. His support for my ideas and intolerance of my disorganization helped make this book happen.

John Berlow contributed beyond measure to the writing of this book. He came into the picture as an unusually intelligent editor. At every phase in the manuscript's development, his critical and enthusiastic readings led to new clarity and new ideas. As the project developed he became, for me, more than an editor. He became a friend, a dialog partner, a critic, and a model of the kind of reader I most want to influence. When I met John he was without computer expertise, although his knowledge in other areas provided him with an immediate base from which to generate his own ideas concerning computers and education.

There are many people whose contributions cannot be categorized. Nicholas Negroponte is a constant source of inspiration, in part precisely because he defies categorization. I also wish to thank Susan Hartnett, Androula Henriques, Barbel Inhelder, A.R. Jonckheere, Duncan Stuart Linney, Alan Papert, Dona Strauss and I.B. Tabata. And there are a few people with whom disagreements about how computers should be used have always been valuable: John Seeley Brown, Ira Goldstein, Robert Davis, Arthur Leuhrman, Patrick Suppes. If the book can be read as an expression of positive and optimistic thinking this must be attributed to my mother, Betty Papert. Artemis Papert has helped in so many ways that I can only say: *Merci.*

Everyone concerned with how children think has an immense general debt to Jean Piaget. I have a special debt as well. If Piaget had not intervened in my life I would now be a "real mathematician" instead of being whatever it is that I have become. Piaget invested a lot of energy and a lot of faith in me. I hope that he will recognize what I have contributed to the world of children as being in the spirit of his life enterprise.

I left Geneva enormously inspired by Piaget's image of the child, particularly by his idea that children learn so much without being taught. But I was also enormously frustrated by how little he could tell us about how to create conditions for more knowledge to be acquired by children through this marvelous process of "Piagetian learning." I saw the popular idea of designing a "Piagetian Curriculum" as standing Piaget on his head: Piaget is par excellence the theorist of learning without curriculum. As a consequence, I began to formulate two ideas that run through this book: (1) significant change in patterns of intellectual development will come about through cultural change, and (2) the most likely bearer of potentially relevant cultural change in the near future is the increasingly pervasive computer presence. Although these perspectives had informed the LOGO project from its inception, for a long time I could not see how to give them a theoretical framework.

I was helped in this, as in many other ways, by my wife Sherry Turkle. Without her, this book could not have been written. Ideas borrowed from Sherry turned out to be missing links in my attempts to develop ways of thinking about computers and cultures. Sherry is a sociologist whose particular concerns center on the interaction of ideas and culture formation, in particular how complexes of ideas are adopted by and articulated throughout cultural groups. When I met her she had recently completed an investigation of a new French psychoanalytic culture, of how psychoanalysis had "colonized" France, a country that had fiercely resisted Freudian influence. She had turned her attention to computer cultures and was thinking about how people's relationships with computation influence their language, their ideas about politics, and their views of themselves. Listening to her talk about both projects helped me to formulate my own approach and to achieve a sufficient sense of closure in my ideas to embark on this writing project.

Over the years Sherry has given me every kind of support. When the writing would not work out she gave me hours of conversation and editorial help. But her support was most decisive on the many occasions when I fell out of love with the book or when my confidence in my resolution to write it flagged. Then, her commitment to the project kept it alive and her love for me helped me find my way back to being in love with the work.

SEYMOUR PAPERT
Cambridge, Massachusetts
April 1980

Notes

Introduction

1. Piaget is at the center of the concerns of this book. I make a slightly unorthodox interpretation of his theoretical position and a very unorthodox interpretation of the implications of his theory for education. The reader who would like to return to the source needs some guidance because Piaget has written a large number of books, most of which discuss particular aspects of children's development, assuming that the others have been read as a theoretical preface. The best short book about Piaget is M. Boden's *Piaget* (London: Harvester Press, 1979). A good starting place for reading Piaget's own texts is with H. E. Gruber and J. J. Voneche, eds., *The Essential Piaget: An Interpretive Reference and Guide* (New York: Basic Books, 1977). My own "short list" of books by Piaget that are most readable and provide the best philosophical overview of his ideas are: *The Child's Conception of the World* (New York: Harcourt, Brace and Co., 1929); *The Child's Conception of Physical Causality* (New York: Harcourt, Brace and Co., 1932); *The Psychology of Intelligence*, trans. Malcolm Piercy and D. E. Berlyne (New York: Harcourt, Brace and Co., 1950); *The Origins of Intelligence in Children*, trans. Margaret Cook (London: Routledge and Kegan Paul); *Introduction à l'Epistémologie Génétique* (Paris: Presses Universitaires de France, 1950); *Insights and Illusions in Philosophy*, trans. Wolfe Mays (New York: The World Publishing Co., 1971); *The Grasp of Consciousness*, trans. Susan Wedgwood (Cambridge: Harvard University Press, 1976). For a critique of the "Piaget Curriculum Developers," of whom I have said that they are "standing Piaget on his head," see G. Groen, "The Theoretical Ideas of Piaget and Educational Practice," *The Impact of Research on Education*, ed. P. Suppes (Washington D. C.: The National Academy of Education, 1978).

2. LOGO is the name of a philosophy of education in a growing family of computer languages that goes with it. Characteristic features of the LOGO family of languages include procedural definitions with local variables to permit recursion. Thus, in LOGO it is possible to define new commands and functions which then can be used exactly like primitive ones. LOGO is an interpretive language. This means that it can be used interactively. The modern LOGO systems have full list structure, that is to say, the language can operate on lists whose members can themselves be lists, lists of lists, and so forth.

Some versions have elements of parallel processing and of message passing in order to facilitate graphics programming. An example of a powerful use of list structure is the representation of LOGO procedures themselves as lists of lists so that LOGO procedures can construct, modify, and run other LOGO procedures. Thus LOGO is not a "toy," a language only for children. The examples of simple uses of LOGO in this book do however illustrate some ways in which LOGO is special in that it is designed to provide very early and easy entry routes into programming for beginners with no prior mathematical knowledge. The subset of LOGO containing Turtle commands, the most used "entry route" for beginners, is referred to in this book as "TURTLE TALK" to take account of the fact that other computer languages, for example SMALLTALK and PASCAL, have implemented Turtles on their systems using commands originally developed in the LOGO language. The TURTLE TALK subset of LOGO is easily transportable to other languages.

It should be carefully remembered that LOGO is never conceived as a final product or offered as "the definitive language." Here I present it as a sample to show that something better is possible.

217

Notes

Precisely because LOGO is not a toy, but a powerful computer language, it requires considerably larger memory than less powerful languages such as BASIC. This has meant that until recently LOGO was only to be implemented on relatively large computers. With the lowering cost of memory this situation is rapidly changing. As this book goes to press, prototypes of LOGO systems are running on a 48K Apple II system and on a TI 99/4 with extended memory. Readers who would like to be kept informed of the status of LOGO implementations can write to me at LOGO project, MIT Artificial Intelligence Laboratory, 545 Technology Square, Cambridge, Mass. 02139. See S. Papert et al., *LOGO: A Language For Learning* (Morristown, N.J.: Creative Computing Press, forthcoming, Summer 1981).

3. The history of the Turtle in the LOGO project is as follows. In 1968–1969, the first class of twelve "average" seventh-grade students at the Muzzy Junior High School in Lexington, Massachusetts, worked with LOGO through the whole school year in place of their normal mathematics curriculum. At that time the LOGO system had no graphics. The students wrote programs that could translate English to "Pig Latin," programs that could play games of strategy, and programs to generate concrete poetry. This was the first confirmation that LOGO was a learnable language for computer "novices." However, I wanted to see the demonstration extended to fifth graders, third graders, and ultimately to preschool children. It seemed obvious that even if the LOGO language was learnable at these ages, the programming topics would not be. I proposed the Turtle as a programming domain that could be interesting to people at all ages. This expectation has subsequently been borne out by experience, and the Turtle as a learning device has been widely adopted. Pioneer work in using the Turtle to teach very young children was done by Radia Perlman who demonstrated, while she was a student at MIT, that four-year-old children could learn to control mechanical Turtles. Cynthia Solomon used screen Turtles in the first demonstration that first graders could learn to program. At the other end of the age spectrum, it is encouraging to see that Turtle programming is being used at a college level to teach PASCAL. See Kenneth L. Bowles, *Problem Solving Using PASCAL* (New York: Springer-Verlag, 1977). Controlling Turtles has proven to be an engaging activity for retarded children, for autistic children, and for children with a variety of "learning disorders." See for example, Paul Goldenberg, *Special Technology for Special Children* (Baltimore: University Park Press, 1979). Turtles have been incorporated into the SMALLTALK computer system at the Xerox Palo Alto Research Center. See Alan Kay and Adele Goldberg, "Personal Dynamic Media" (Palo Alto, Calif.: Xerox, Palo Alto Research Center, 1976).

4. *Touch Sensor Turtle*. The simplest touch sensor program in LOGO is as follows:

	Comments
TO BOUNCE	
REPEAT	This means repeat all the individual steps
FORWARD 1	The turtle keeps moving
TEST FRONT.TOUCH	It checks whether it has run into something
IFTRUE RIGHT 180	If so, it does an about turn
END	

This will make the Turtle turn about when it encounters an object. A more subtle and more instructive program using the Touch Sensor Turtle is as follows:

	Comments
TO FOLLOW	
REPEAT	
FORWARD 1	
TEST LEFT.TOUCH	Check: Is it touching?
IFTRUE RIGHT 1	It thinks it's too close and turns away
IFFALSE LEFT 1	It thinks it might lose the object so it turns toward
END	

218

This program will cause the Turtle to circumnavigate an object of any shape, provided that it starts with its left side in contact with the object (and provided that the object and any irregularities in its contour are large compared to the Turtle).

It is a very instructive project for a group of students to develop this (or an equivalent) program from first principles by acting out how they think they would use touch to get around an object and by translating their strategies into Turtle commands.

Chapter 1

1. The program FOLLOW (See Introduction, note 4) is a very simple example of how a powerful cybernetic idea (control by negative feedback) can be used to elucidate a biological or psychological phenomenon. Simple as it is, the example helps bridge the gap between physical models of "causal mechanism" and psychological phenomenon such as "purpose."

Theoretical psychologists have used more complex programs in the same spirit to construct models of practically every known psychological phenomenon. A bold formulation of the spirit of such inquiry is found in Herbert A. Simon, *Sciences of the Artificial* (Cambridge: MIT Press, 1969).

2. The critics and skeptics referred to here are distillations from years of public and private debates. These attitudes are widely held, but, unfortunately, seldom published and therefore seldom discussed with any semblance of rigor. One critic who has set a good example by publishing his views is Joseph Weizenbaum in *Computer Power and Human Reason: From Judgment to Calculations* (San Francisco: W.H. Freeman, 1976).

Unfortunately Weizenbaum's book discusses two separate (though related) questions: whether computers harm the way people think and whether computers themselves can think. Most critical reviews of Weizenbaum have focused on the latter question, on which he joins company with Hubert L. Dreyfus, *What Computers Can't Do: A Critique of Artificial Reason* (New York: Harper & Row, 1972).

A lively description of some of the principal participants in the debate about whether computers can or cannot think is found in Pamela McCorduck, *Machines Who Think* (San Francisco: W.H. Freeman, 1979).

There is little published data on whether computers actually affect how people think. This question is being studied presently by S. Turkle.

3. Many versions of BASIC would allow a program to produce a shape like that made by the LOGO program HOUSE. The simplest example would look something like this:

```
10 PLOT (0,0)
20 PLOT (100,0)
30 PLOT (100,100)
40 PLOT (75,150)
50 PLOT (0,100)
60 PLOT (0,0)
70 END
```

Writing such a program falls short of the LOGO program as a beginning programming experience in many ways. It demands more of the beginner, in particular, it demands knowledge of cartesian coordinates. This demand would be less serious if the program, once written, could become a powerful tool for other projects. The LOGO programs SQ, TRI, and HOUSE can be used to draw squares, triangles, and houses in any position and orientation on the screen. The BASIC program allows one particular house to be drawn in one position. In order to make a BASIC program that will draw houses in many positions, it is necessary to use algebraic variables as in PLOT (x, y), PLOT $(x + 100, y)$, and so on. As for defining new commands, such as SQ, TRI, and HOUSE, the commonly used versions of BASIC either do not allow this at all or, at best, allow something akin to it to be achieved through the

Notes

use of advanced technical programming methods. Advocates of BASIC might reply that: (1) these objections refer only to a beginner's experience and (2) these deficiencies of BASIC could be fixed. The first argument is simply not true: The intellectual and practical primitivity of BASIC extends all along the line up to the most advanced programming. The second misses the point of my complaint. Of course one could turn BASIC into LOGO or SMALL-TALK or anything else and still call it "BASIC." My complaint is that what is being foisted on the education world has not been so "fixed." Moreover, doing so would be a little like "remodelling" a wooden house to become a skyscraper.

Chapter 2

1. "Gedanken experiments" have played an important role in science, particularly in physics. These experiments would encourage more critical attitudes if used more often in thinking about education.

2. There is a joke here. Readers who are not familiar with Noam Chomsky's recent work may not get it. Noam Chomsky believes that we have a language acquisition device. I don't: the MAD seems no more improbable than the LAD. See N. Chomsky, *Reflections on Language* (New York: Pantheon, 1976) for his view of the brain as made up of specialized neurological organs matched to specific intellectual functions. I think that the fundamental question for the future of education is not whether the brain is "a general purpose computer" or a collection of specialized devices, but whether our intellectual functions are reducible in a one-to-one fashion to neurologically given structures.

It seems to be beyond doubt that the brain has numerous inborn "gadgets." But surely these "gadgets" are much more primitive than is suggested by names like LAD and MAD. I see learning language or learning mathematics as harnessing to this purpose numerous "gadgets" whose original purpose bears no resemblance to the complex intellectual functions they come to serve.

Chapter 3

1. Since this book is written for readers who may not know much mathematics, references to specific mathematics are as restrained as possible. The following remarks will flesh out the discussion for mathematically sophisticated readers.

The isomorphism of different Turtle systems is one of many examples of "advanced" mathematical ideas that come up in Turtle geometry in forms that are both *concrete* and *useful*. Among these, concepts from "calculus" are especially important.

Example 1: Integration. Turtle geometry prepares the way for the concept of line integral by the frequent occurrence of situations where the Turtle has to integrate some quantity as it goes along. Often the first case encountered by children comes from the need to have the Turtle keep track of how much it has turned or of how far it has moved. An excellent Turtle project is simulating tropisms that would cause an animal to seek such conditions as warmth, or light, or nutrient concentration represented as a field in the form of a numerical function of position. It is natural to think of comparing two algorithms by integrating the field quantity along the Turtle's path. A simple version is achieved by inserting into a program a single line such as: CALL (:TOTAL + FIELD) "TOTAL", which means: take the quantity previously called "TOTAL," add to it the quantity FIELD and call the result "TOTAL." This version has a "bug" if the steps taken by the Turtle are too large or of variable size. By debugging when such problems are encountered the student moves in a meaningful progression to a more sophisticated concept of integral.

The early introduction of simple version of integration along a path illustrates a frequent phenomenon of reversal of what seemed to be "natural" pedagogic ordering. In the traditional curriculum, line integration is an advanced topic to which students come after having been encouraged for several years to think of the definite integral as the area under a curve, a concept that seemed to be more concrete in a mathematical world of pencil and paper technology. But the effect is to develop a misleading image of integration that leaves many students with a sense of being lost when they encounter integrals for which the representation as area under a curve is quite inappropriate.

Example 2: Differential Equation. "Touch Sensor Turtle" (See introduction, note 4) used a method that strikes many children as excitingly powerful. A typical first approach to programming a Turtle to circumnavigate an object is to measure the object and build its dimensions into the program. Thus if the object is a square with side 150 Turtle steps, the program will include the instruction FORWARD 150. Even if it works (which it usually does not) this approach lacks generality. The program cited in the earlier note works by taking tiny steps that depend only on conditions in the Turtle's immediate vicinity. Instead of the "global" operation FORWARD 150 it uses only "local" operations such as FORWARD 1. In doing so it captures an essential core of the notion of differential equation. I have seen elementary school children who understand clearly why differential equations are the natural form of laws of motion. Here we see another dramatic pedagogic reversal: The power of the differential equation is understood before the analytic formalism of calculus. Much of what is known about Turtle versions of mathematical ideas is brought together in H. Abelson and A. diSessa, *Turtle Geometry: Computation as a Medium for Exploring Mathematics* (Cambridge: MIT Press, in press).

Example 3: Topological Invariant. Let a Turtle crawl around an object "totalizing" its turns as it goes: right turns counting as positive, left turns as negative. The result will be 360° whatever the shape of the object. We shall see that this Total Turtle Trip Theorem is useful as well as wonderful.

2. The phrase "ego-syntonic" is used by Freud. It is a "term used to describe instincts or ideas that are acceptable to the ego: i.e., compatible with the ego's integrity and with its demands." (See J. Laplanche and J-B. Pontalis, *The Language of Psycho-analysis* (New York: Norton, 1973.)

3. G. Polya, *How to Solve It* (Garden City, N.Y.: Doubleday-Anchor, 1954); *Induction and Analogy in Mathematics* (Princeton, N.J.: Princeton University Press, 1954); and *Patterns of Plausible Inference* (Princeton, N.J.: Princeton, 1969).

4. Usual definitions of curvature look more complex but are equivalent to this one. Thus we have another example of an "advanced" concept in graspable form.

5. If turns can be right or left, one direction must be treated as negative. "Boundary of (connected) area" is a simple way of saying "simple closed curve." If the restriction is lifted, the sum of turns must still be an integral multiple of 360.

Chapter 4

1. Here I am picking a little quarrel with Jerry Bruner. But I share much of what he thinks, and this is true not only about language and action, but also about the relationship to learning of cultural materials and of teaching.

The systematic difference between us is seen most clearly by comparing our approaches to mathematics education. Bruner, as a psychologist, takes mathematics as a given entity and considers, in his particular rich way, the processes of teaching it and learning it. I try to *make* a learnable mathematics. I think that something of the same sort separates us in regard to language and to culture and leads us to different paradigms for a "theory of learning." See J.S. Bruner, *Toward a Theory of Instruction* (Cambridge: Harvard University Press, 1966) and J.S. Bruner et al., *Studies in Cognitive Growth* (New York: John Wiley, 1966).

Notes

2. The most systematic study is in H. Austin, "A Computational Theory of Physical Skill" (Ph.D. thesis, MIT, 1976).

3. These procedures introduce a further expansion in our image of programming. They are capable of running simultaneously, "in parallel." An image of programming that fails to include this expansion is quite out of touch with the modern world of computation. And a child who is restricted to serial programming is deprived of a source of practical and of conceptual power. This deprivation is felt as soon as the child tries to introduce motion into a program.

Suppose, for example, that a child wishes to create a movie on the computer screen with three separate moving objects. The "natural" way to do this would be to create a separate procedure for each object and set the three going. "Serial" computer systems force a less logical way to do this. Typically, the motions of each object would be broken up into steps and a procedure created to run a step of each motion in cyclic order.

The example shows two reasons why a computer system for children should allow parallel computation or "multi-processing." First, from an instrumental point of view, multi-processing makes programming complex systems easier and conceptually clearer. Serial programming breaks up procedural entities that ought to have their own integrity. Second, as a model of learning serial programming does something worse: It betrays the principle of modularity and precludes truly structured programming. The child ought to be able to construct each motion separately, try it out, debug it, and know that it will work (or almost work) as a part of the larger system.

Multi-processing is more demanding of computational resources than simple serial processing. None of the computers commonly found in schools and homes are powerful enough to allow it. Early LOGO systems were "purely serial." More recent ones allow restricted forms of multi-processing (such as the WHEN DEMONS described later in this chapter) tailored for purposes of programming dynamic graphics, games, and music. The development of a much less restrictive multi-processing language for children is a major research goal of the MIT LOGO Group at the time of writing this book. In the work we draw heavily on ideas that have been developed in Alan Kay's SMALLTALK language, on Carl Hewitt's concepts of "ACTOR" languages and on the Minsky-Papert "Society Theory of Mind." But the technical problems inherent in such systems are not fully understood and much more research may be needed before a concensus emerges about the right way (or set of ways) to achieve a really good multi-processing system suitable for children.

Chapter 5

1. The most prolific contributor to the development of such systems is Andrea diSessa, who is responsible, among many other things, for the term "Dynaturtle." H. Abelson and A. diSessa, *Turtle Geometry: Computation as a Medium for Exploring Mathematics* (Cambridge: MIT Press, in press).

2. The discussion of the Monkey Problem uses a computational model. However this model is very far from fitting the notion of computation as algorithmic programming built into most programming languages. Making this model consists of creating a collection of objects and setting up interactions between them. This image of computation, which has come to be known as "object-oriented" or "message-passing" programming, was first developed as a technical method for simulation programs and implemented as a language called SI-MULA. Recently it has drawn much broader interest and, in particular, has become a focus of attention in Artificial Intelligence research where it has been most extensively developed by Carl Hewitt and his students. Alan Kay has for a long time been the most active advocate of object-oriented languages in education.

Notes

Chapter 6

1. Martin Gardner, *Mathematical Carnival* (New York: Random House, 1977).

Chapter 7

1. For remarks by Piaget on Bourbaki see "Logique et connaissance scientifique," ed. J. Piaget, *Encyclopidie de la Pleide,* vol. 22 (Paris: Gallimard, 1967).

2. C. Lévi-Strauss, *Structural Anthropology,* 2 vols. (New York: Basic Books, 1963–76).

3. Lévi-Strauss uses the word *bricolage* as a technical term for the tinkering-like process we have been discussing. *Bricoleur* is the word for someone who engages in *bricolage.* These concepts have been developed in a computational context in Robert Lawler, "One Child's Learning: An Intimate Study" (Ph.D. thesis, MIT, 1979).

4. Of course our culture provides everyone with plenty of occasions to *practice* particular systematic procedures. Its poverty is in materials for *thinking about* and *talking about* procedures. When children come to LOGO they often have trouble recognizing a procedure as an entity. Coming to do so, is, in my view, analogous to the process of formation of permanent objects in infancy and of all the Piagetionly-conserved entities such as number, weight, and length. In LOGO, procedures are manipulable entities. They can be named, stored away, retrieved, changed, used as building blocks for superprocedures and analyzed into subprocedures. In this process they are assimilated to schematic or frames of more familiar entities. Thus they acquire the quality of "being entities." They inherit "concreteness." They also inherit specific knowledge.

223

Index

Acceleration Turtles, 128
Aesthetic, mathematical, 190–206
  inborn sense of, 203
  logic and, 204
Aesthetic guidance, 195–198
Aesthetic sense, innate, 190–191
Aesthetics of physics, 122
Agents:
  competing, 172
  modular, 173
  theory of, 167–169, 170, 173
AI, *see* Artificial intelligence
Algebraic mother structures, 160
Analytic geometry, 97, 98–99
Analytic versus intuitive selves, 97
Angle, concept of, 68
Anthropomorphism, 59
Aristotle, 121, 141–144, 166
Articulateness, development of mathematical, 75–76
Articulation:
  debugging and, 180
  importance of ability of, 158
Artificial intelligence (AI), 157–158
  case frame thinking in, 199
  concretizing quality of, 156–157
  deductive systems and, 195
  defined, 157
  epistemological modularity and, 171
  "object-oriented" or "message-oriented" programming and, 222n 2 (Ch. 1)
  as research methodology, 164–166
Assimilation, Piaget's concept of, *vii,* 120
Austin, Howard, 111–113

BASIC, 29, 33–36, 218n 2
  example of program in, 219–220n 3
  pidgin English and, 211
"Being wrong," 23, 62, 101, 114
Bernouilli's law, 165
Biological sciences, 68
Body geometry, 56
Body knowledge, *viii,* 9
Body syntonicity, 63, 68, 205
Bourbaki school structuralism, 159–160, 207
  roots of mathematics and, 163–164
Bricolage/bricoleur, 173, 175, 223n 3

Bruner, Jerome, S., classification of ways of knowing, 96, 221n 1 (Ch. 4)
Bugs, 22, 23, 101–102; *see also* Debugging
  in complex process, 112
  in juggling, 111
  in learning physical skills, 104
  multiple, 112–113

Calculus, 66–67, 221n 1 (Ch. 3)
Cartesian geometry, 55, 66–67, 97, 98–99
Child as builder of intellectual structures, 7
  incentives and, 22–23
  materials and, 19, 22–23
  of mathematics, preconservationist and conservationist, 126, 129
  of microworlds, 118–119, 162
  teaching without curriculum and, 32
  of theories, 132–134, 172
Child as epistemologist, 19, 23, 27–28, 98
Child-computer relationship, 19
  conditions necessary for, 16
Combinatorial thinking, 21–22, 174–176, 197
Computation, samba schools for, 182–183
Computational-method theory of thinking, 167–169
Computational model for people procedures, 106–107
Computational theory of agents, 173
Computer(s), 19–37
  child's relationship to, 5–6
  cost of, 17, 23–25, 34
  editing with, 13, 31
  effect of, on children, 29–30
  human mind and, 26–27
  as instrument for drill and practice, 36, 139
  as instrument of change, 209
  as mediator of relationships, 183–184
  natural communication with, 6–7
  as object-to-think-with, 23
  as pencil, 210
  personal, 24, 181
  in personal lives, 32
  simulative power of, *viii*
  society and, 25–26
  as teaching instrument, 30–31

225

Index

Index

Index

29930185R00143

Made in the USA
Lexington, KY
12 February 2014